Lavender Farming

A Step by Step Guide to Lavender Farming and It's Benefits

(Guide to Growing Lavender Plants for Massive Profit and Everything You Need to Know About the Lavender Farming)

Nicholas Miller

Published By **Elena Holly**

Nicholas Miller

All Rights Reserved

Lavender Farming: A Step by Step Guide to Lavender Farming and It's Benefits (Guide to Growing Lavender Plants for Massive Profit and Everything You Need to Know About the Lavender Farming)

ISBN 978-1-7774403-0-5

No part of this guidebook shall be reproduced in any form without permission in writing from the publisher except in the case of brief quotations embodied in critical articles or reviews.

Legal & Disclaimer

The information contained in this book is not designed to replace or take the place of any form of medicine or professional medical advice. The information in this book has been provided for educational & entertainment purposes only.

The information contained in this book has been compiled from sources deemed reliable, and it is accurate to the best of the Author's knowledge; however, the Author cannot guarantee its accuracy and validity and cannot be held liable for any errors or omissions. Changes are periodically made to this book. You must consult your doctor or get professional medical advice before using any of the suggested remedies, techniques, or information in this book.

Upon using the information contained in this book, you agree to hold harmless the Author from and against any damages, costs, and expenses, including any legal fees potentially resulting from the application of any of the information provided by this guide. This disclaimer applies to any damages or injury caused by the use and application, whether directly or indirectly, of any advice or information presented, whether for breach of contract, tort, negligence, personal injury, criminal intent, or under any other cause of action.

You agree to accept all risks of using the information presented inside this book. You need to consult a professional medical practitioner in order to ensure you are both able and healthy enough to participate in this program.

Table Of Contents

Chapter 1: Farming With Lavender A Aromatic Journey 1

Chapter 2: Choosing The Right Lavender Varieties For Your Farm 6

Chapter 3: Understanding Lavender Plant Anatomy And Growth Patterns 11

Chapter 4: Preparing The Soil For Successful Lavender Cultivation 16

Chapter 5: Essential Tools And Equipment For Lavender Farming 21

Chapter 6: Propagation Techniques Growing Lavender From Seeds As Well As Cuttings 26

Chapter 7: Nurturing Lavender Seedlings 32

Chapter 8: Best Practices For Lavender Field Layout And Plant Spacing 38

Chapter 9: Irrigation Methods For Optimal Lavender Watering 45

Chapter 10: Nutrient Management 51

Chapter 11: Controlling Weeds, Pests, And Diseases In Lavender Farming 59

Chapter 12: Pruning And Shaping Lavender Plants For Health And Aesthetics 65

Chapter 13: Harvesting Lavender 73

Chapter 14: Post-Harvest Handling: Drying And Storing Lavender 80

Chapter 15: Identifying Value-Added Products Derived From Lavender 86

Chapter 16: Lavender Oil Extraction 92

Chapter 17: Crafting Lavender-Based Cosmetics And Skincare Products 102

Chapter 18: Lavender In Culinary Delights 110

Chapter 19: Examining The Medicinal Properties Of Lavender 120

Chapter 20: Creating Lavender Sachets And Potpourri 127

Chapter 21: Lavender Farming For Essential Oil Production 133

Chapter 22: Marketing And Selling Your Lavender Products 142

Chapter 23: The Fascinating History Of Lavender 150

Chapter 24: Why Lavender Is A Profitable Crop 153

Chapter 25: Choosing The Right Variety 157

Chapter 26: Growing Lavender 161

Chapter 27: Harvesting And Drying Lavender 164

Chapter 28: Lavender Products 167

Chapter 29: Marketing Your Lavender Business 170

Chapter 30: The Wellness Benefits Of Lavender 173

Chapter 31: Lavender As A Home Decorator To Create Soothing Spaces The Scent Of This Flower 176

Chapter 32: Advanced Lavender Techniques 179

Chapter 1: Farming With Lavender A Aromatic Journey

Lavender farming is an enthralling and rewarding undertaking that can provide both aromatic satisfaction and the potential for economic gains. In addition to its enchanting scent, the many uses it can be used for that lavender is a favorite in the imagination of many lovers gardening enthusiasts, entrepreneurs, and gardeners across the globe. In this section we take an exciting journey through the intricate details of farming lavender and explore its past as well as the promising future it offers for the modern farming methods.

Section 1: The Enchanting History of Lavender

Lavender's tale begins a long time ago that is the result of antiquated civilizations like that of Egyptians, Greeks, and Romans. Its usage dates back to the Pharaohs and Egyptians, who used the herb for cosmetics and embalming. The Romans appreciated lavender's beneficial properties for relaxation and utilized to bathe in it, and the Greeks believed in its medicinal properties. The widespread cultivation of lavender can be traced back to the Romans who brought lavender to a variety of regions in Europe, Asia, and North Africa.

Section 2: Lavender's Botanical Marvels

In order to fully comprehend the the cultivation of lavender, it's essential to understand the botanical features of this amazing plant. The lavender plant is part of the Lamiaceae family as well as the genus of Lavandula that covers more than 47 species. Lavandula angustifolia is commonly referred to in the name English lavender is one of the most sought-after species in commercial cultivation because of its unique scent as well as its oil concentration. Lavender plants are famous for their slim silvery-gray leaves and vivid red, purple or white blooms that draw pollinators and make an ideal feature in any garden.

Section 3: Lavender's Versatile Uses

The enthralling scent of lavender and its numerous uses make it an incredibly diverse plant with enormous commercial possibilities. From its aromatic oils to medicinal and culinary uses the lavender plant offers an variety of options. Essential oils extracted from the lavender plant is highly sought-after in the world of fragrance, which is used in soaps, perfumes candles, soaps, and many other products that are scented. Also, lavender has made its place in the kitchen to enhance dishes by adding aromatic and herbal taste. Additionally, the therapeutic qualities of lavender are a major reason why it is a sought-after component in aromatherapy skin care products and natural cures.

Section 4: The Benefits of Lavender Farming

Lavender farming can bring many advantages. Beyond the aesthetic and olfactory delights it provides it can also become a profitable venture. The market for products made from lavender is on the rise due to the increasing desire for natural and sustainable alternative options. The lavender farms are able to tap into different revenue streams like the selling of dried or fresh lavender bundles, products made from lavender such as essential oils and Agritourism opportunities. In addition, the cultivation of lavender promotes environmentally sustainable practices in agriculture, since it's a resilient plant which thrives in dry environments and has minimum chemical inputs.

Section 5: The Challenges and Rewards of Lavender Farming

Although lavender farming is an exciting opportunity but it's crucial to understand the risks it poses. It is sensitive to water amounts, conditions of the soil and climate conditions, requiring the careful selection of the right site and cultivation practices. Furthermore, the presence of pests, disease and competition with weeds can create significant problems for lavender growers. But with the right understanding, planning and the implementation of good techniques, these hurdles are able to be overtaken. The benefits of farming lavender like the enjoyment of working in the natural world and the beautiful lavender fields, as well as the pleasure of making high-end items, are far greater than the difficulties.

Section 6: Embarking on Your Lavender Farming Journey

For a successful the lavender farming process It is essential to build a solid base of understanding and expertise. The book guides through each step of the process, starting from choosing the appropriate varieties of lavender and then preparing the soil, to harvesting the plants, processing them, and marketing your products made from lavender. The book will explore the intricate aspects of growing lavender cultivation, post-harvest, and other methods, providing your with the essential tools to start and run the success of your lavender farm.

Chapter 2: Choosing The Right Lavender Varieties For Your Farm

Opting for the best lavender cultivars is an essential step to creating a productive lavender farm. There is a broad selection of cultivars and species that are available, each with specific characteristics and adaptability it is essential to take care to determine the most suitable one for the farm's climate the soil, demands of the market and individual preferences. In this article we'll explore aspects to take into consideration when selecting the best lavender variety, look at popular cultivars and provide guidelines to assist you make educated choices.

Section 1: Understanding Lavender Species and Cultivars

The Lavandula genus which includes many species and cultivars. The two species most commonly used to be grown commercially include Lavandula angustifolia (English lavender) as well as Lavandula x intermedia (lavandin). English lavender is famous for its unique scent as well as its oil quality, which makes it an ideal choice for cooking scent therapy, cosmetic, and aromatherapy reasons. Lavandin is a mix of English lavender, and the spike variety (Lavandula latifolia) is a better yielding plant but a less appealing oil quality.

Section 2: Things to consider when selecting lavender Variety

Many factors can determine your choice when choosing lavender to plant on your farm:

Climate: The varieties of lavender show different degrees of cold-hardiness in addition to heat tolerance and resistance to drought. Be aware of the climate in your area and pick varieties that are suitable for your area's climate and rain patterns.

Soil Conditions: The lavender thrives in alkaline, well-drained soil. Check your soil's pH levels the texture, as well as fertility levels to identify the kinds of plants that thrive within your soil's specific conditions.

Goal and Demand for Markets Find out the objective of your lavender farm whether for dried or fresh bundles of lavender or essential oils, food applications, or even added value items. Examine the market demand for particular lavender items within your region to help guide the variety of options you choose.

The size of the plant and its growth habits Variety of Lavenders differ in their size and patterns, from small and bushy, to tall and spreading. Take into consideration the space available on your property and the aesthetics you want and maintenance ease when choosing the varieties.

Section 3: Popular Lavender Varieties for Commercial Cultivation

There are a variety of lavender types offered, each of which has its own features and benefits. These are the most popular ones which are widely used to be used for commercial use:

Lavandula angustifolia 'Munstead' Compact English lavender plant with fragrant purple flowers. It's renowned for its toughness and its early flowering time so it is suitable for colder climates.

Lavandula angustifolia "Hidcote' Another English lavender cultivar featuring deep purple flowers and the scent is strong. "Hidcote" is loved by its decorative value as well as the quality of its oil.

Lavandula angustifolia "Grosso": Lavandin varieties that produce tall flower spikes that have vibrant shades of purple. "Grosso" is prized due to its large oil yield and is commonly used in the production of commercial oils.

Lavandula x intermedia "Provence" The lavandin plant is famous for its stunning blue-violet blooms and powerful fragrance. "Provence" is widely utilized in dried arrangements of flowers as well as extraction of essential oils.

Lavandula Stoechas 'Anouk' Also known for its Spanish lavender. The cultivar features unique flowers that resemble butterflies and sweet fragrance. "Anouk" is admired because of its small size as well as its resistance to drought and heat.

Section 4: Local and Regional Considerations

If you are considering lavender plants It is crucial to take into consideration local and regional influences that could affect their growth. Get advice from local lavender growers or agricultural extension service or experts in horticulture who are familiar with growing lavender in your locale. They will provide important information about what varieties have proved to be successful in the specific microclimate of your area and can provide guidance specific to the specific conditions of your area.

Section 5: Trial and Error

The process of experimentation is essential in determining the best lavender cultivars to plant on your farm. Begin with a smaller test plot and planting various kinds to see their growth and growth patterns, as well as flower quality and oil content. A trial and error approach lets the gardener to get firsthand information and help you make educated decisions about the future plantations.

Chapter 3: Understanding Lavender Plant Anatomy And Growth Patterns

In order to cultivate the lavender successfully for a long time, you must be aware of the anatomy of the plant and development patterns. If you are familiar with the anatomy and developmental phases of the lavender plant to improve your farming methods to make sure that you are ensuring the health and longevity of your lavender harvest. In this section we'll dive into the intricate details of anatomy of the plant. We will examine its development patterns and provide useful information to help you through the different stages of cultivation.

Section 1: Lavender Plant Structure

The lavender plant has a distinct design that is characterized by its slender leaf, silvery-gray leaves and vivid blooms. We will look at the various aspects of a lavender plant:

Roots: The plants of the lavender develop an elongated root system which anchors them to the soil. They absorb moisture and nutrients. They spread out horizontally looking for water and creating a solid foundations that the plant can rely on.

The stem the lavender plant is slender and square, adorned with the scent of aromatic leaves. It is structurally strong and transports water, nutrients as well as hormones between roots as well as the remainder parts of the plants.

Leaves: The leaves of lavender are slender, long and needle-like. They are laid out in opposing pairs along the stems. They're covered in tiny hairs that contribute to its distinctive silver appearance and aid in reducing water loss by transpiration.

Flowers: The lavender flowers represent the crowning jewels of the plant. They adorn the stems in vivid hues and enchanting scent. They are usually planted in thick spikes that are composed of multiple blossoms that contain reproduction structures that are essential for the production of seeds.

Section 2: Lavender Growth Patterns

Understanding the patterns of growth in lavender is essential for proper control and treatment. Let's look at the different phases of growth in lavender:

The Seedling and Germination Stages Seedlings of lavender sprout under good conditions, generally in a properly-constructed seedbed or in a container. Seedlings emerge, with an initial cotyledon-like leaves that are to be followed by development of leaves. At this point the most important thing is to provide sufficient humidity, warmth as well as protection from harsh temperatures.

Vegetative Growth: When lavender plants grow older and mature, they begin the stage of vegetative growth. In this phase they focus on establishing a sturdy root system as well as increasing their leaf. Plants that produce lavender sprout new leaves that develop into stems and branches and result in a stronger and stronger the plant.

Flowering Stage: The stage of flowering is a visually and thrillingly satisfying phase of the development of lavender. Plants that grow lavender produce buds that slowly open to reveal vibrant and fragrant blooms. The butterflies, bees, as well as other pollinators, are drawn to the blooms, which aids with cross-pollination as well as the production of seeds. Careful attention during this phase like the management of irrigation and pest control is crucial to guarantee an optimal production of flowers as well as quality.

Seed Production: Following a the successful pollination process, lavender flowers turn into seed pods, which are that are referred to as capsules. They contain tiny, brown seeds, which will be dispersed after they mature. If you want to increase the production of seeds it is important for the capsules to fully mature and dry out on the plant prior to harvesting.

Section 3: Factors Influencing Lavender Growth

A variety of factors impact the development and development of the lavender plant. Knowing these influences will allow you to provide optimal conditions to your lavender plants:

Lighting: The lavender thrives in the sun's full rays, and requires minimum six to eight hours of direct sunshine every day to ensure optimal development and bloom production. A lack of sunlight can lead to fragile plants that have elongated stems as well as fewer flowers.

Soil: Lavender thrives in the well-drained, slightly alkaline soil that has good fertility. Insufficient drainage could cause root rot as well as various other ailments, and the soil's excessive richness can lead to an excessive growth of the leaves which can affect bloom production. Tests of soils and appropriate modifications can provide the perfect growing environment for lavender.

It is drought-tolerant after it is established. However, it will require regular irrigation during the first growth phases and during periods of long-lasting drought. Insufficient watering could lead to root rot as well as other fungal ailments, which is why it's crucial to establish the proper balance, and to avoid excessive watering.

Chapter 4: Preparing The Soil For Successful Lavender Cultivation

The basis of a profitable lavender farm rests on the cultivation of soil. Making sure that the soil is in a good condition that allows lavender plants to flourish is crucial to their health, development as well as productivity. In this article we'll explore the various steps to prepare the soil for a successful lavender farming, which includes soil testing, amendments to the soil and the correct soil management strategies. When you understand the importance of the preparation of soil, you will be able to prepare the ground for an enviable lavender farm.

Section 1: Importance of Soil Preparation

A proper soil preparation is vital for the cultivation of lavender due to these reasons:

drainage: The plants of lavender require drainage that is well-maintained to avoid flooding, which could lead to root rot, among various other illnesses. When you make sure that the soil drains properly ensures a healthy conditions for growing lavender.

Access to Nutrients: Proper nutrition is vital for healthy development as well as development for lavender plant. Preparing soil helps increase the amount of nutrients available, while ensuring that the necessary elements are available in adequate quantity.

pH Balance Lavender likes alkaline soil that has an pH range from 6.5 up to 8.5. Soil preparation lets you alter the pH of the soil to your ideal range to ensure optimal intake of nutrients and minimising nutrient deficits.

The proper soil preparation can reduce the growth of weeds and reduces the competition for resources, and decreasing the use of herbicides.

Section 2: Soil Testing

Prior to beginning the lavender cultivation it is essential to conduct a soil test. It is essential to determine the soil's nutrients and pH. The results of soil testing provide valuable data that can help guide the amendment process. Methods used in soil testing are:

For collecting soil samples, use an auger for soil or a shovel for collecting soil samples at multiple places in your lavender fields. Make sure you collect samples from a similar depth of 6-8 inches and make sure that every sample reflects the normal conditions of the soil in the region.

Submitting Soil Samples the soil samples collected in well-maintained containers, clearly marking them with the appropriate sampling area. Send the soil samples to a trusted soil testing lab for analysis.

How to Interpret Soil Test Results When you have received the report of your soil test, look over the results to better understand the pH and nutrient levels of the soil. It will provide guidelines for amending your soil in accordance with the requirements of lavender.

Section 3: Soil Amendment

The process of soil amendment is adding organic matter as well as adjusting the pH of the soil and levels of nutrients in order to provide the best conditions for the growth of lavender. The exact amendments needed will vary based on test results of the soil. Be aware of these practices:

Organic Matter: The incorporation of organic matter, like compost that is well-rotted or used manure, helps improve soil drainage, structure as well as nutrient retention. Sprinkle a layer or organic matter across the surface of your soil and then incorporate it in the top 1-2 inches of soil with the tiller or a garden fork.

pH Adjustment: In the event that the soil pH is excessively acidic, adding lime from agriculture could raise the pH up until it is within the range desired. In contrast, if soil is acidic mineral sulfur, or other acidic substances could aid in lowering the pH. Use the guidelines within the soil analysis report, and include the amendments in the soil with care.

Nutrient Amendments based on soil tests results, use the suggested fertilizers or amendments to the soil to correct any deficiencies in nutrient levels. It is common for lavender to benefit from balanced fertilizers that have low nitrogen levels, to limit an excessive growth of foliage, which can be which can affect the production of flowers.

Mulching: Spread a blanket of mulch made from organic materials for example, straw, or even wood chip on the lavender plants. Mulching can help retain moisture in the soil as well as stifle weed growth and helps regulate the temperature of soil.

Section 4: Soil Management Techniques

When the soil amendments have been integrated, the proper soil control techniques will maintain the performance and health for your farm's lavender. Be aware of these practices:

The plants of lavender require regular irrigation, especially in the growth phases that begin. In the end, too much water can result in damage. Install an irrigation system that gives consistent water without overwatering the soil.

Chapter 5: Essential Tools And Equipment For Lavender Farming

The cultivation of lavender requires a variety of equipment and tools for an efficient harvest, cultivation, and post-harvest treatment. When you choose the best machinery and devices, you'll be able to simplify your operation, increase production, and boost the overall efficiency of your farm. In this section we'll look at the most essential equipment and tools essential to grow lavender that cover the entire process from field prep through the post harvest handling.

Section 1: Field Preparation Tools

Tiller, or circular cultivator is vital to prepare the soil prior to plant lavender. It assists in breaking down soil that is compacted, make amendments and make a fertile and well-aerated plant bed.

Rake: A strong metal rake is ideal for making sure that the soil is level as well as removing dirt as well as creating a level seedbed.

Spade or Shovel: A spade or shovel is required to dig holes for planting as well as moving soil and transfer of compost and other organic materials.

Soil Testing Kit soil testing kit will allow you to determine the amount of nutrients and the pH levels of your soil. This will assist you to make an informed decision regarding amending your soil.

Section 2: Planting and Propagation Tools

Pruning Shears: Pruning scissors often referred to secateurs, are vital for cutting lavender to remove dead or diseased leaves, as well as shaping plant for optimum development and aesthetics.

Seed Tray, also known as a Propagation Tray: The seed tray, also known as a propagation tray, is employed to plant the seeds of lavender or for cuttings to root. It is a secure setting for seed germination as well as root development.

Seedling Transplanter transplanter, like dibbers or tools for transplanting can help create holes for planting for delicate lavender seedlings, without damaging the roots of their plants.

Plant labels: You can use labels or markers to mark the different varieties of lavender or plant areas in your fields. This is crucial if you grow many varieties of lavender.

Section 3: Irrigation Tools

Drip Irrigation System A drip irrigation system is the most effective method of providing the lavender plant with water. It provides the water directly to the plant the root zones, thus reducing losses of water as well as reducing the chance of developing foliar diseases.

Hose and Sprinkler Hose equipped with a sprinkler attachment could be used to provide supplemental irrigation particularly in the initial phases of the growth of lavender. It gives a wide area of coverage, and can be useful in setting up new plants.

Section 4: Harvesting Tools

Harvesting Shears: Harvesting Shears often referred to harvest knives, or sickles are utilized to cut laurels' stems in the harvest. Pick shears with an edge that is serrated for smooth cuts that cause minimal harm to the lavender plant.

Harvest baskets or totes Get sturdy containers or totes to store harvesting lavender bundles. Opt for materials with breathable properties which allow for air circulation in order to avoid moisture accumulation.

Section 5: Post-Harvest Handling Equipment

Drying Rack Drying Rack drying table is necessary in air drying lavender bundles harvested from the field. It offers ample space to hang the lavender bundles, and also allows an adequate airflow that will aid in drying.

Stripping Tray: A stripping tray, also known as a stripping board can be employed to take dried lavender buds off the stems. It is made up of a screen or mesh area on which it is possible to gently rub stems and allow the flowers to drop into the container beneath.

Essential oil distillation equipment If you intend to make lavender essential oils, then specialized distillation equipment, like steam distillers as well as an essential oil distiller, is necessary. The equipment lets you draw essential oils from lavender blooms.

Section 6: General Farm Tools and Equipment

Wheelbarrows or Garden Cart Wheelbarrows or garden cart can be useful in transportation of soil amendments, harvesting lavender, equipment, as well as other items all around the farm.

A Pruning tool is essential for more extensive pruning projects, for example cutting off thick lavender branches, or reviving plants that have been in decline.

Your Personal Protective Equipment (PPE): Be sure to wear the appropriate PPE that includes protective eyewear, gloves and sturdy footwear for your protection while working in the field.

Chapter 6: Propagation Techniques Growing Lavender From Seeds As Well As Cuttings

Propagating lavender is an important ability for farmers of lavender which allows them to increase the lavender field, preserve desirable varieties and provide the availability of plants that are healthy. There are two main methods of propagation for lavender: starting with cuttings or seeds, and also from cuttings. In this section we'll explore both techniques more in-depth, offering detailed step-by-step instruction and helpful tips that will help you propagate lavender to achieve you want to achieve.

Section 1 The Growing of Lavender From Seeds

The cultivation of lavender seeds is a cheap way to multiply a lot of plant species. But, it takes dedication and patience. It is a process that requires patience and attention to detail:

Seed Collection: Harvest seeds of mature lavender plants once the flowers are become brown and dried. The flowers can be gently rubbed through your fingers until you allow the seeds to release their small brown ones. Keep the seeds and place them in a dry container.

Seed Stratification: The seeds of lavender are benefited by a period of stratification under cold temperatures to end the dormancy of their seeds and boost the rate of germination. The seeds should be placed in an absorbent tissue or seed-starting mixture then seal them with the bag of a plastic container, and put them in a refrigerator for 2 - four weeks.

Seed Starting Mix: Create an initial mix for seeds consisting of equal portions of perlite, peat moss vermiculite, and perlite. This light, well-drained mix provide an ideal substrate for the germination of lavender seeds.

Seed Sowing: fill the seed containers or trays with mixture for seeding. Plant the lavender seeds onto the top of the mix by press them gently downwards. Overlay the seeds with a light layer of vermiculite, or seed mix.

Maintenance and watering: Moisten the seed starting mix with a fine spray. Make sure it is constantly damp but not too wet. The trays should be placed in a sunny location that receives indirect light. The trays can be covered with transparent plastic domes or plastic wrap to make the illusion of a mini greenhouse. It will also keep moisture in.

The process of germination and transplanting is similar to that for lavender. seeds usually germinate in about 2 to 3 weeks. When the seedlings are able to develop their leaves, cautiously transfer them to small pots or seedling trays that are filled with potting soil well-drained. Make sure you provide adequate lighting, and slowly adjust the seedlings to outdoor conditions prior to transferring them to the field.

Part 2: Growing lavender From Cuttings Growing lavender from cuttings is a proven method of maintaining specific lavender cultivars with desirable characteristics. This is how it works:

Picking the Right Cuttings: Select healthy, disease-free lavender flowers that have strong growth, and attractive qualities for cuttings. Pick stems that do not flower. They are semiripe. That means they're neither soft and neither too hard. Ideally, stems should measure three to five inches.

Making the Cuttings: With the sharp, clean pruning shears Take cuttings from chosen stems, just below the leaf node. Cut off the lower leaves and leave a few leafs at the uppermost. Cut the cut into a length of about 2-3 inches at the level of the leaf node.

The hormone called rooting: to encourage the growth of roots development Dip the cut edge of your lavender cutting into powder for rooting or gel. The manufacturer's directions will guide you on the right amount of hormone you should use.

Growing the Cuttings: Make the rooting medium by mixing equally parts of vermiculite and perlite or using a well-drained potting mix. Make small pots or trays with the soil and create holes with the dibber or pencil. Incorporate the roots into the holes and by gently settling the soil on top of the cuttings.

Disperse and cover: Spray the plants with water in order to bring the soil down over the plants. Put a transparent plastic bag or lid for a propagator over the trays or pots in order for a moist environment. The water will remain moist and promotes root development.

Transplanting and Rooting The cuttings of lavender usually root in between 4 and six weeks. Be sure to monitor the cuttings frequently and ensure that the medium for rooting is not dripping wet, but it's not inundated with water. When the roots have grown slowly acclimatize the cuttings outdoors before transferring them to the field.

Section 3: Care and Maintenance of Propagated Plants

Following the propagation of lavender cuttings or seeds It is vital to provide adequate care and upkeep to guarantee the health of the plants you have propagated. Think about the following aspects:

The seedlings should be watered or roots cuttings frequently to ensure the soil is uniformly damp. Do not overwater as lavender plants can suffer from root rot in humid conditions.

Transplanting: Once the cuttings, seedlings or roots are at the right size and risk of frost is gone then transplant them to the field. Make sure that the plants are spaced evenly in order to permit adequate airflow and avoid crowding.

Weed Control: Use ways to manage weeds for example, mulching or hand-weeding to reduce competition with weeds and to ensure that there is no weed growth surrounding your lavender plant propagation.

Fertilization: The plants of the lavender require just a little fertilization. Utilize a slow-release, balanced fertilizer, or add organic substances when preparing the soil in order to provide essential nutrients.

Chapter 7: Nurturing Lavender Seedlings

Growing lavender seedlings is an vital stage in the cultivation of lavender which requires care and proper care in order to guarantee the health of their development. The lavender seedlings are fragile and vulnerable. They require special conditions and methods for growth before they're at the point of transplanting to the field. In this section we'll look at the most important ways to care for and transplant lavender seedlings. We will also provide guidance regarding the care of these plants, and preparing them to be successful.

Section 1: Germination Care

A proper care plan in the time of germination sets the stage for healthful lavender seeds to grow. Be sure to follow these guidelines:

The temperature and light conditions The best time to germinate lavender seeds is in temperatures between 60degF to 70 degF (15degC between 21 and 20degC). Make sure you have a warm and light-filled area to allow seed germination making sure that the seed trays are exposed to indirect sunlight, or have supplementary light sources such as fluorescent.

Moisture Management: Maintain a consistent levels of moisture in seed trays in the germination process. Spray the soil often to keep it damp but not waterlogged. Make use of the fine mist sprayer in order to prevent seeds from dispersing or creating compaction in the soil.

Air circulation: A proper air circulation is essential to avoid fungal infections. The dome is removed or cover every now and then to permit fresh air to circulate, however ensure that you maintain adequate water levels.

Section 2: Seedling Care

When the lavender seedlings appear Continue to provide the attention that they require to develop into strong plant. Take note of the following tips:

Lighting and Temperature: Put seeds in an area in which they get plenty of light. Move them into a cooler climate with temperatures that range from 55degF-65degF (13degC up to

18degC) to promote strong and dense development.

The seedlings should be watered cautiously to ensure a constant amount of moisture in the roots. Make use of a watering container fitted with a fine rose or a mist sprayer provide an easy, uniform distribution of water. Do not overwater as too much humidity can result in root mold.

Fertilization: Seedlings of the lavender plant are not in need of a heavy dose of fertilizer in this phase. Start applying a dilute, well-balanced fertilizer, like an 10-10-10 or 5-5-5 formulation after the seeds develop their first real leaves. Be sure to follow the instructions of the manufacturer for the correct dilution amount and the frequency at which you apply.

Thinning: When multiple seedlings appear in a container, you can slim them down by eliminating those with weaker seeds, and permitting the stronger ones to expand without competing. Pick the best and healthiest robust seedlings for that you have the highest chance of a successful outcome.

Section 3: Transplantation Preparation

The preparation of lavender seedlings to be transplanted is the most important step in their progress towards being field-ready plants. Take these steps to ensure an easy transition

The Seedlings are Hardened Off The seedlings should be hardened off by slowly acclimatizing them to the outdoor environment. Start by placing them outside in a protected area for just a couple of hours every day. Gradually increase the time as well as exposed to sunlight throughout the span between 1 and two weeks. This will help the seeds adapt to the outdoors and decreases the chance of suffering from transplant shock.

Transplantation Time: Plant lavender seeds into the field on the day following the last date of frost in your area after the soil has been warmed up and the risk of frost is over. It is usually mid-summer or late summer. Choose a day that has moderate temperatures and clear skies in order to reduce stress on the seeds.

Soil Preparation: Create the holes to plant in the soil prior to transplanting the seeds. Make sure the soil is properly drained as well as fertilized with organic matter and free of the weeds. The holes should be positioned according to the ideal spacing of your particular lavender plant, which is usually between 2 and three feet from each other.

Section 4: Transplantation Techniques

Utilize these methods to ensure a successful transfer of lavender seedlings

Digging Holes: Create holes into the area slightly bigger than the roots of seeds. It is important to ensure that they are wide enough to allow for the roots, without stretching or overcrowded.

Carefully take the seedlings out of their container, making sure not to harm the roots which are delicate. Keep the seedling in place by the leaf or root ball Avoid excessive manipulation of the stem.

Then, backfill and place seeds into the holes. Ensure that the upper part of the rootball is at or just over the soil's surface. Fill the holes back with soil by gently firming the soil around the roots in order in order to get rid of air pockets.

When transplanting is completed, immediately after the it is essential to water your seedlings well to help settle the soil, and aid in the establishment of roots. Make sure you provide enough water so that it can saturate the root zone, but not create the conditions of waterlogging.

Mulching: Spread a thin blanket of mulch made from organic material to the seeds' base including straw, wood chippings or even straw. This can help retain moisture as well as smother weeds and control the temperature of the soil.

Section 5: Post-Transplantation Care

The proper care following transplantation of lavender seedlings is essential to ensure their proper growth on the field. Use these tips for post-transplant care:

Chapter 8: Best Practices For Lavender Field Layout And Plant Spacing

The design of your field for lavender and the distance between your plants play an essential part in your overall health, productivity as well as the aesthetics of your farm. A well-planned layout of your field and proper spacing of plants will ensure proper sun exposure, air circulation and effective farm managing practices. In this section we'll explore the best practices in lavender field layout as well as spacing of plants, and provide advice for maximising the benefits for your lavender harvest.

Section 1: Field Layout Considerations

Be aware of the following aspects in determining the design of your lavender field:

The sun's rays thrive in the sun's full rays, and needs minimum six to eight hours of direct sunshine per day. Pick a place where your lavender plant has plenty of sunlight during the day. Be sure to avoid zones that are shaded by the trees or other structures.

Soil Drainage: The lavender plants require well-drained soils to stop the root from rotting and waterlogging. Pick a spot with draining soil, or use soil preparation methods like mounding or raised beds, for proper drainage.

Topography: Select an area that is flat or lightly sloped lavender garden to allow for the efficient irrigation process and to prevent soil erosion. Slopes that are steep can cause problems when it comes to water management. They also make it more likely for erosion in the soil.

Accessibility: Think about the ease of access to your lavender field to farming management activities including the planting, irrigation, pruning and harvest. Make sure there's enough space for machines and equipment to move around without difficulty.

Section 2: Plant Spacing Guidelines

A proper plant spacing is vital in maximizing the growth of airflow, as well as overall well-being of lavender plants. Take note of the following tips for determining spacing between plants:

Varieties to Consider: The varieties of lavender can differ in their growing habits, size and strength. Certain varieties naturally have larger, more compact size some may spread or spread further. Be aware of the particular characteristics of the lavender you're cultivating before determining the spacing of your plants.

Spacing between Rows: Leave ample space between rows in order for equipment, staff movements, as well as airflow. An ideal guideline is to ensure a gap between 3 and four feet (0.9 to 1.2 meters) between rows. This will ensure that sufficient space is available for harvesting and maintenance.

Spacing Between Rows: Spacing between each lavender plant in a row is determined by the type of lavender and functional and aesthetic features. Here are some suggestions for spacing basing on popular lavender varieties:

Lavandula angustifolia The plant English lavender varieties about 2-3 inches (0.6 up to 0.9 meters) from each other in rows. The spacing allows enough space for plants to develop and permits enough airflow as well as sunlight to penetrate.

Lavandula Intermedia (lavandin): Lavandin varieties are typically bigger and stronger in comparison to English lavender. The lavender varieties are planted 3-4 inches (0.9 up to 1.2 meters) from each other in rows in order to allow for their growing habits and ensure good air flow.

The hedge or border Plantings If you intend to plant borders or hedges made of lavender You can place the plants closer to create an appearance that is more dense. To create hedges, place lavender plants 2 to 1 inches (0.3 or 0.6 meters) from each other in rows so as for lateral growth. This will make a thick, even hedge.

Section 3: Benefits of Proper Plant Spacing

Proper plant spacing offers numerous benefits for your lavender farm:

Air Circulation: Proper spacing permits the proper circulation of air between plants, which reduces the possibility of fungal illnesses and improving overall health of plants. Airflow that is efficient also assists to pollinate by making sure there is a smooth flow of pollinators.

Sunlight Exposure: Proper spacing of plants ensures that every lavender plant gets enough sunlight to promote robust growth bloom production and oil development.

Space for Weed Control: A proper spacing allows for easier management of the weeds that grow between rows as well as within the plants of lavender. A good spacing will allow for adequate access for manual removal of weeds and to implement mulching methods effectively.

Easy Maintenance: Correct plant spacing makes it easier to perform tasks of maintenance, like pruning, harvesting and control of pests. The space between plants is sufficient the efficient movement of plants and minimizes the chance of damaging neighbouring plants in the course of these chores.

Aesthetics: Lavender plants that are well-placed make for a pleasing and even field. A proper spacing improves the aesthetics and beauty of your lavender farm and makes the area visually appealing to both visitors as well as customers.

Section 4: Field Management Considerations

As well as ensuring an appropriate spacing of your plants, think about these field management techniques to increase the effectiveness for your lavender plants:

irrigation: Use an efficient irrigation system, for example drip irrigation, which delivers fresh water to the plant's root zone. The proper irrigation management is vital for the lavender's needs in terms of water and helps prevent water waste.

Weed Control: Implement effective methods to control weeds, like hand-weeding, mulching or using approved herbicides in order to limit competition with weeds, and to maintain an untidy field. The management of weeds is especially important in the initial phase of lavender plants.

Pruning and training: Regularly cut and train lavender plants so that they maintain their shape, stimulate the growth of more bushy plants, as well as encourage maximum flower production. Pruning reduces plant size increases airflow and lowers the chance of contracting disease.

Soil Health: Check and ensure the health of your soil by adding organic matter for example, compost or well-rotted manure during the preparation of soil. A regular soil test can to ensure that the soil has adequate nutrients and equilibrium of pH.

Chapter 9: Irrigation Methods For Optimal Lavender Watering

The proper use of irrigation is vital to the growth and health of the lavender plant. The lavender plant requires a balanced amount of water for its growth since both watering underwater or overwatering could have negative consequences for its growth as well as general health. In this article we'll explore different ways of irrigation to get the best drainage for lavender. This will ensure that your plants are getting enough moisture to meet their particular needs.

Section 1: Understanding Lavender's Water Requirements

Before diving into the methods of irrigation is essential to know the requirements for water in lavender. Take note of the following aspects:

Soil Moisture: Lavender flourishes in soil that is well-drained and favors a moderately dry or slightly humid soil. A lot of moisture, or soil that is waterlogged could cause root rot as well as other illnesses.

Drought Tolerance Lavender is well-known for its drought-tolerant qualities when it is established. In the event of overwatering, it can lead to reduced essential oil production, reduced quality of the flower, as well as an increase in risk of contracting diseases.

The stage of growth in which lavender's requirements for water are different throughout the stage of growth. Although young lavender plants need regular watering, mature plants that have strong root systems are able to withstand longer time periods with no irrigation.

Section 2: Irrigation Methods for Lavender

Select the best irrigation technique depending on the specific growth conditions as well as the requirements of your particular lavender crop. These are the commonly-used methods of irrigation for the most effective lavender irrigation:

Drip Irrigation The drip irrigation method is the most preferred method of lavender cultivation since it gives specific and reliable water supply.

It involves utilizing tubes that has drippers or emitters placed close to the bottom of every lavender plant. Drip irrigation directs water into the plant's root zone to reduce water loss and decreasing the chance of getting foliar diseases. It's especially efficient for saving water in areas with the limited availability of water.

Soaker Hoses: Soaker-type hoses are an alternative method of lavender irrigation. They have small pores, which let water be slowly absorbed into the soil and ensures an extensive root system permeation. Install the soaker hoses in the lines of lavender plants and allow them to supply water directly to the roots. Soaker hoses can be particularly beneficial to plant new lavender seeds or transplants.

Sprinkler Irrigation The use of sprinklers can be employed as a secondary irrigation strategy, particularly when the plant is beginning to establish lavender plants. Overhead sprinklers provide large coverage and distribute water in tiny drops over the field of lavender. But, care must be employed to prevent wetting the plants too often, as they can cause foliar disease. When using sprinkler irrigation It is recommended to soak lavender plants in the morning so that the foliage to dry prior to sunset.

Hand Watering: Hand-watering is an efficient method, especially for small lavender planters or container-grown. Make use of a watering container fitted with a delicate rose attachment to bring water to the roots of the plants. This will prevent overly wetting the leaves. Make sure you water the plants thoroughly and ensure your soil remains well-watered at the root level without leaving waterlogged areas.

Section 3: Irrigation Frequency and Timing

Determining the frequency and duration of irrigation is essential for maintaining proper soil water levels for lavender. Follow these guidelines:

Monitoring of Soil Moisture: Regularly check the moisture of your soil for determining when irrigation is required. Utilize a moisture meter for soil or conduct a finger test. Simply insert your fingers into the soil around the roots of the plant. If you feel that the soil is dry, to the extent of around two inches (5 centimeters) this indicates that irrigation is required.

In general, lavender plants require irrigation after the surface one to two inches (2.5 to 5 centimeters) of soil is evaporated. But, it is important to avoid watering excessively, as this could cause weak root development. Once the plants are mature and develop deeper root systems decrease the frequency of watering to allow the roots to search for more moisture in the soil.

The best time to water lavender is at the beginning of the day is advised. This will allow the plants to absorb the amount of water they require in the midst of the hot day. Any excess water on the leaves has an opportunity to dry out prior to sunset, which reduces the chance of developing foliar diseases. Do not water in later afternoon or evening since prolonged periods of humidity can contribute to diseases development.

Section 4: Irrigation Management Considerations

As well as selecting the appropriate timing and method of irrigation be aware of the following elements to ensure you are managing the irrigation of your lavender plant:

Mulching: Place a thin covering of organic mulch for example, straw or wooden chips at the base of lavender plants. Mulching aids in retaining soil moisture and reduce weed growth and helps regulate the temperature of soil. It also reduces evaporation. This reduces the requirement for regular irrigation.

Evapotranspiration: Take note of the rate of evapotranspiration in your area, which refers to the loss of water by transpiration of the soil's surface and transpiration through the leaves of plants. Look up local agricultural resources or weather stations to assess the evapotranspiration rates and modify your irrigation methods accordingly.

Chapter 10: Nutrient Management

A proper management of nutrients is crucial for the growth as well as the productivity of the lavender plant. Although lavender is renowned as a plant that thrives under poor soil conditions providing the correct level of nutrients could dramatically improve the vigor of your lavender plants as well as general quality. In this section we'll explore the significance of managing nutrient levels to grow lavender and talk about the proper use of fertilizers and soil amendments in order to improve the quality of your lavender crops.

Section 1: Understanding Lavender's Nutrient Requirements

There are specific requirements for nutrient intakes in lavender to help support its development and development. Take note of the essential nutrients that lavender plants require:

Nitrogen (N) It is vital for the promotion of the growth of plants as well as leaf development as well as general plant strength. But, too much nitrogen could cause excessive growth of the foliage and decrease the production of flowers and the amount of essential oils.

Phosphorus (P) is a key component of Phosphorus. It is a key element in the root development as well as flower development as well as overall energy transfer in the plant. Particularly important is the beginning phases of the growth process and the establishment.

Potassium (K): Potassium helps to improve overall health of the plant in disease prevention, as well as control of the flow of water throughout the plant. It is a key ingredient in flowering quality and essential oil production and stress tolerance of plants.

Secondary Nutrients: The plant also needs secondary nutrients such as calcium (Ca) and magnesium (Mg) as well as sulfur (S) that are necessary for the proper growth of plants. development, activation of enzymes, and the uptake of nutrients.

Micronutrients: Micronutrients including iron (Fe) manganese (Mn) and zinc (Zn) and copper (Cu) as well as Boron (B) as well as molybdenum

(Mo) are essential in small amounts however they are essential to perform various metabolic functions in the lavender plant.

Section 2: Soil Testing and Nutrient Analysis

When applying fertilizers and soil amendments, it's important to determine the nutritional status of the soil. Testing your soil and analysing nutrient levels provide important information regarding the soil's nutrients and pH. Follow these steps:

Soil Sampling: Take the soil samples of the lavender field with an auger for soil or soil sampling device. You can collect samples from several spots, but avoid areas that are not typical like fences or zones that were heavily affected with fertilizer treatments in the past.

Laboratory Analysis: Take the soil samples to an accredited laboratory for soil testing to be analysed. The lab will provide an in-depth report on the soil's nutritional levels along with recommendations to apply fertilizers based on the nutrient needs of your lavender.

pH Adjustment Lavender is a fan of a somewhat acidic or neutral pH between 6.0 to 7.5. If the pH of your soil is not within this range, your soil test report may recommend the adjustment of pH. Lime is usually employed to increase the pH of soil, but acidic or sulfur-based amendments could decrease it.

Section 3: Fertilizers for Lavender

Fertilizers may be utilized to meet the nutritional requirements for lavender plants. Follow the guidelines below when the application of fertilizers:

Organic Fertilizers: Organic fertilisers, including compost, manure that has been well-rotted, and organic amendments provide slow release nutrients, and help increase soil structure as well as moisture retention. They also improve the overall wellbeing of the soil's ecosystem. Use organic fertilizers in the preparation of soil or for top dressings on the bases of the lavender plant.

Inorganic Fertilizers: Organic or synthetic fertilizers may be utilized to provide specific nutrients as they are required. Select balanced fertilizers or ones specifically formulated for lavender, with ratios of N-P-K suitable for the requirements of lavender's nutrients. Refer to the directions of the manufacturer's on applications rates and application times.

Application Time: Apply fertilizers to lavender plants at right times to ensure optimal nutrients intake. In general, lavender plants benefit from applying a small amount of fertilizer during the spring months just before the new growth starts as well as after the initial blooming cycle, to help support bloom production as well as the essential oils development. Beware of applying fertilizers later in the season of growth, because it could hinder the flower's winter period of dormancy.

Section 4 section 4: Amendments to Soil Amendments to Lavender Soil amendments can enhance the soil's structure, fertility and capacity for water retention and benefit the lavender plant. Take note of the following changes:

Organic Matter: Add organic matter decomposed well including compost into the soil prior to plant lavender. Organic matter enhances the soil's structure, improves the accessibility of nutrients as well as promotes beneficial microbe activity.

Perlite and vermiculite: Both perlite and vermiculite are light mineral amendments that help improve the aeration of soils and drainage. By incorporating these amendments into soil or mixing them in with the soil prior to preparation may aid in creating well-drained soil suitable for lavender.

Gypsum: Gypsum is a calcium sulfate amendment which can help improve drainage and structure of soil in clay-based soils that are heavy. Include gypsum in the soil in the process of soil preparation, to facilitate greater water circulation and lessen soil compaction.

Epsom Salt Epsom salt (also known as magnesium sulfate) can be employed in soil amendments to provide an additional source of sulfur and magnesium. Magnesium is vital for chlorophyll growth, while sulfur is helpful in nutrient absorption as well as plant health. Use Epsom salt in a controlled manner and adhere to the prescribed rates for application.

Section 5: Nutrient Management and Environmental Considerations

The proper management of nutrients should be mindful of environmental concerns as well as sustainable practices.

Environmental Impact: Apply fertilisers and amendments to soils in a controlled manner according to the recommended rate of application. In excess, nutrients could cause water pollution, runoff from nutrient-rich soils, and even the destruction of the environment. Use the best practices for managing to limit the loss of nutrients and safeguard waters resources.

Organic and sustainable practices Implement environmentally sustainable and organic practices in your lavender cultivation. Make sure you are building healthier soil ecosystems, maximizing the cycling of nutrients, while minimizing the usage of synthetic inputs. Use crop rotation, cover crops as well as composting to increase the fertility of soil organically.

Monitor regularly the appearance and growth of the lavender plant in order to determine their needs for nutrients. Signs of the visual world, like foliage color, rate of growth and overall plant strength could indicate deficiencies in nutrient levels or excessive amounts. Modify your management of nutrient requirements in line with the signs.

Chapter 11: Controlling Weeds, Pests, And Diseases In Lavender Farming

Effective weed, pest and disease management is essential to ensure the health, effectiveness, and attractiveness of farms producing lavender. They can be a source of competition and encroach on the lavender plant, while pests could cause damage to flowers and foliage, while diseases could weaken or could even kill lavender plants. In this article we'll look at strategies and the best methods for controlling the spread of pests, weeds and diseases that affect lavender farming and ensuring the health of your lavender plant.

Section 1: Weed Control in Lavender Farms

The weeds are a quick issue in farms with lavender plants as they compete with the lavender plant for sunlight, nutrients and water. Below are a few effective methods to control weeds:

Mulching: Place a thin covering of organic mulch including straw or wood chip or pine needles, at the lavender plant's root. Mulching can help reduce the growth of weeds because it blocks sunlight, keeping soil moisture, as well as reducing seeds of weeds germination. Make sure the mulch layer is dense enough to stop weeds from entering.

Hand Weeding: Frequently inspect the lavender plants in your garden and eliminate any weeds you see. Hand weeding can be particularly beneficial in removing large or persistent unwanted weeds that cannot be easily controlled through mulching on their own. Make sure to be cautious when hand-weeding so as to not damage the plants' weak roots.

Cultivation: Utilize tools for cultivation like hoes or hand cultivators to alter the surface of soil and stop the growth of weeds. Planting the soil between the rows of lavender will assist in controlling weeds and breaking the root system. Make sure not to disturb the plants' small roots while cultivating.

Weed Barrier Fabric Use an weed barrier or landscape fabric to surround lavender plants. These fabrics make a physical barrier which keeps weeds out and helps conserve the soil's water. Make sure the fabric is held securely and has sufficient coverage of each lavender plant.

Pre-emergent Pre-emergent herbicides can be applied prior to seeds germinate to stop the growth of weeds. Use herbicides with labels that indicate their use in farms that grow lavender, be sure to follow the directions of the producer attentively. Important to remember that pre-emergent herbicides can inhibit the growth of plants you want to cultivate Therefore, be cautious and use only when necessary.

Section 2: Pest Control in Lavender Farms

Pests can do a lot of damages to the lavender plant and can affect the growth of their the flowers, their foliage, as well as general well-being. Here are a few strategies to ensure successful pest control for lavender farms:

Integrative Pest Control (IPM) Use an IPM strategy for pest control. It is focused on prevention including cultural practices, biological control before taking chemical solutions. Monitor your plants regularly to look for signs of pests. immediately take action if you notice symptoms of pest invasion.

Beneficial Insects: Provide the existence of beneficial insects like ladybugs, lacewings and parasitic wasps that feed on commonly-found lavender bugs like aphids as well as Thrips. You can plant nectar-rich blooms such as daisies and yarrow to draw beneficial insect species to your lavender garden.

Companion Planting: Mix lavender alongside companion plants that naturally deter bugs or draw beneficial insects. Like, for example, putting marigolds, catnip or chives in conjunction with lavender may assist in preventing pests from entering and improve the effectiveness of pest control.

Organic Pest Control: Make use of organic methods for pest control like insecticidal soaps, or the oils for gardening, to manage insects on the lavender plant. They are not than harmful to beneficial insects, and do not have a negative impact on the ecosystem. Be sure to follow the instructions of the manufacturer for appropriate application and usage.

Pest Exclusion: Guard your young lavender plants against pests employing physical barriers, such as rows covers or nets. They can stop animals, like deer and rabbits from gaining access to the lavender plant and causing damage.

Section 3: Disease Management in Lavender Farms

Plant diseases can harm lavender which can cause leaves spots, wilting and reduced growth and dying plants. To manage the spread of diseases within lavender farms, think about these methods:

The importance of plant health and cultural practices Keep your plants healthy through providing appropriate growing conditions such as well-drained soils with adequate sun exposure as well as the proper spacing. A good cultural practice like appropriate irrigation and avoidance of excessive overhead irrigation, will aid in reducing the incidence of disease.

Cleanliness: Keep your home clean to stop spreading diseases. Get rid of any affected plants or plant matter that is diseased as soon as it becomes apparent. Cleaning pruning tools in between the plants to prevent transmission of disease.

Types of Disease-resistant Variety: Choose disease-resistant lavender plants whenever possible. Some varieties of lavender possess better protection against common ailments which reduces the risk of contracting.

Fungicides: If needed you must use fungicides that are specially designed for lavender farms to combat fungal infections. Be sure to follow the instructions of the manufacturer and suggested rates of application. It is crucial to keep in mind that chemical controls should only be utilized in the last instance and only when all other methods of preventive and cultural measures are not working.

Chapter 12: Pruning And Shaping Lavender Plants For Health And Aesthetics

Pruning and shaping the lavender plant is an essential part of lavender cultivation to maintain healthy plants, promoting strong expansion, improving bloom production and enhancing attractiveness. Correct pruning methods help preserve the size and shape of the plants. They also improve circulation of air, stop disease as well as encourage the development of fragrant, essential oils-rich blooms. In this section we'll discuss the significance of pruning as well as shaping the plants of lavender and examine the best methods to get the most effective outcomes.

Section 1: Understanding the Benefits of Pruning Lavender Plants

The lavender plant's pruning can provide a variety of advantages to their performance, health, and general appearance. Take note of the following benefits:

Pruning can stimulate lateral branching and results in a larger lavender plant. It creates a larger form, which is more compact and increases the overall densities.

Improves Flower Growth Pruning encourages development of the lateral buds which produce flowering plants. When you remove the blooms that are no longer needed and trimming back the plant will stimulate the growth of flower spikes that are new that will lead to a prolonged and greater quantity of blooms.

Enhances the air Circulation Pruning allows air circulation to the inside of the lavender plant, enhancing the flow of air and lessening the possibility of fungal disease. A good airflow reduces the amount of moisture retained on the foliage, and creates the health of plants.

Protects against Disease Pruning eliminates plants that are diseased or dead which reduces the chance of fungal infection as well as bacterial infections. Pruning allows better sunlight access and quicker drying of leaves, which creates undesirable conditions for pathogens.

Pruning plants shape Pruning is a way to maintain the desirable shape and size of lavender plants. Pruning allows you to make borders, hedges or plants that are neatly designed which enhance the overall appearance of the lavender farm you have.

Section 2: When to Prune Lavender Plants

It is important to be punctual in pruning lavender plants. Take note of the following tips for proper pruning

Spring Pruning: Plants of lavender are best pruned during the early spring when new growth begins to emerge. The timing of pruning allows the plant to heal after pruning, and then produce fresh plants just in time for blooming time.

Post-Bloom Pruning After the initial period of bloom it's beneficial to trim the plant lightly to eliminate flower spikes that are no longer in use. This helps in the development of fresh flower spikes and also extends the blooming period.

A light summer pruning: For areas that have long-lasting summers, light summer pruning is a good idea to form lavender plants and stop the growth from becoming excessive. But, it is important not to prune to late in the time, because it can hinder the plant's winter period of dormancy.

Section 3: Pruning Techniques for Lavender Plants

Use these pruning methods for optimal results when cutting lavender plants:

Cleanse Pruning Tools Before pruning, clean the pruning equipment including pruning shears and hedge trimmers using an alcohol solution comprised that contains 70% alcohol or disinfectant. This can help stop the spread of disease among plants.

Take out the spent flowers: Following the first blooming period get rid of the flower spikes and cut them below a pair or healthy buds or leaves. Clean cuts are the best way to limit the chance of introducing disease.

Light Pruning to Shape to maintain a tidy form, gently prune leaves on the edges of lavender plants. Cut back any loose or overgrown branches. Cut the branches just over the healthy buds or leaves.

Do not cut in Woody Stems: When pruning lavender, be careful not to cut into the stems that are woody or cutting off all of the green growth. The woody bottom is what helps the plant maintain its structure, and promotes the growth of healthy plants.

Gradual Renovation Pruning: With time the lavender plants turn woody and less productive. For a rejuvenation of these plants, use the gradual renovation pruning strategy. Reduce by one-third the height of the plant each year, over three years. This will allow the plant to undergo gradual renewal without causing shock to the plant.

Section 4: Shaping Lavender Plants

Shaped lavender plants can enhance their beauty and creates consistency in the lavender farm. The following guidelines will help you shape lavender plants:

Regular Shearing: When you prefer a professional or manicured style, it is recommended to regularly cut your lavender plants so that they maintain an elongated, compact shape. Make use of sharp hedge trimmers or shears to make precise cuts at the top of a healthy collection of leaves.

Natural Form: If you like natural and freeform style, only minimal shaping is necessary. Simple pruning to keep the plant in good well-being and eliminate dead or damaged development is sufficient.

Border or Hedge Formation For lavender borders or hedges cut the plants to the same size and height. Make use of stakes and strings for guideposts to create smooth and consistent lines.

Tapered Shape: To create the shape of a tapered, trim the lavender plants to ensure they're a little wider at the base, and more narrow toward the top. The shape improves the structure of the plant and also provides attractive design.

Section 5: Pruning Tips and Considerations

Check out the additional suggestions and tips when pruning lavender plants:

Pruning is important, you should not over-prune lavender plants because it may result in decreased blooms and weaker development. Only prune lightly as necessary to ensure the health of the plant and its shape.

Watch for signs of stress Watch your plants' responses to pruning. If you see indications of stress, like wilting, or slow growth, you should reevaluate your pruning techniques and modify according to the need.

Pruning Frequency: Regularly pruning is essential to ensure the condition and form that lavender flowers take. Plan for a regular pruning program in springtime as well as occasional light pruning throughout the growing season to get rid of the blooms that have died and to maintain their its shape.

Beware of pruning in cold Weather Avoid Pruning during Cold Weather to not prune lavender plants in colder weather, or in the event of a chance of frost. Pruning at these times could cause the plant to be more susceptible to freezing damage.

Get rid of pruned materials Pruning is the proper way to dispose of pruned lavender plant material to stop spreading of disease or bugs. Get rid of pruned materials from the garden or compost it with care.

Pay attention to the specific needs of each plant The individual lavender plants may be unique in its growth pattern and demands. Examine the pattern of growth of the individual plant and modify pruning strategies accordingly, to ensure the uniform and even appearance.

Chapter 13: Harvesting Lavender

The right time and using appropriate methods is essential to guarantee best quality lavender blossoms and essential oil. If you harvest too early or late may result in lower results or less fragrance. In this section we'll explore the ideal timing to harvest lavender, the methods employed, as well as tools needed to ensure a fruitful harvest when you are working on your lavender farm.

Section 1: Determining the Right Time for Harvesting Lavender

It is vital to be punctual for harvesting lavender to ensure the highest scent and quality. Take into consideration the following aspects in determining the ideal time to harvest:

Blooming Stage: Lavender needs to be picked when flowers are at their peak blooming phase. The optimal time to pick is at the time that about half the flowers on the stem have started to open while the rest are about to open. This is when the blooms are at their best for fragrance as well as essential oils.

Colour and Appearance: Lavender flowers must display vibrant colors and an attractive appearance. They must be well-developed and not show any evidence of browning or witting. Be careful not to harvest lavender in the time of or immediately after rain as the excessive moisture may affect the flower's quality.

The best time to harvest lavender early in the morning, after the dew has dried, but ahead of the scorching heat of the day. In this time the oil content of the flower is the highest making it the strongest scent.

Section 2: Techniques for Harvesting Lavender

Use the following methods to ensure that you get the best harvest of lavender:

Hand Harvesting: When you have small lavender farms, or when you are looking for high-quality using hand-harvesting methods is the most effective technique. Be sure to hold the stalk of the lavender plant near the spikes of flowers and then use a sharp shear or scissors to cut an exact cut. Keep a few inches of stem tied to the blooms to make it easier for you to handle the flowers after harvest processing.

Bundling: Gather lavender into smaller bundles that are easier to handle and drying. Collect around 100 to 150 stems making sure the flower spikes are aligned as well as adjusting the length of the stems to create consistency. Attach the bundle using twine or a rubber band and leave a length of trail for hanging.

Continuous Harvesting: Farms of lavender that have a lot of plants could decide to continue harvesting. Instead of harvesting the entire lavender simultaneously, it is better to select plants to harvest when they get to the desired bloom stage. This permits an even harvest and ensures an ongoing supply of new blooms all through the flowering season.

Pruning for Harvest: If you are picking lavender, you should consider pruning methods that are specifically targeted to harvest. In other words, leave lavender plants that have some leaves after harvesting, to encourage growth and ensure that the plant is healthy.

Section 3: Essential Tools for Harvesting Lavender

Use the tools below to help you harvest your crops and make sure that your operations are efficient

Pruning Scissors or Shears Buy the most precise, quality pruning shears or scissors specially designed for use in horticulture. They should feature an efficient and clean cutting technique to ensure that they don't damage the flowers and stems.

Twine or rubber bands Attach lavender bouquets using twine or rubber bands to hold their shape, and to prevent spikes of flower from falling down in drying. Select items that are safe for use in the agricultural sector and won't cause harm to lavender.

Apron for Harvesting or Basket: You can use the harvesting apron, or basket to store the lavender stems that have been cut in the process of harvesting. The stems are kept in order and helps prevent damage or bruising on the flowers.

Protective Gear: Put on suitable protective equipment like gloves or long sleeves to shield your arms and hands from abrasions and possible irritation due to the lavender leaves.

Section 4: Post-Harvest Processing

The proper post-harvest treatment is crucial for preserving the aroma and quality of the harvested lavender. Take note of the following actions:

Hanging and Bundling: hang the lavender bouquet upside-down in a dark, cool well-ventilated space. Find a place in which the bundles are safe from intense sunlight and temperatures. The lavender will dry completely, until the stems have a crisp appearance and the flowers keep their scent and color.

Destemming and storage: When the lavender bouquets have completely dried, you can gently remove the bouquets of flowers by cutting off the stems. Keep the dried flowers in glass or airtight containers in a cool, dark location to keep their fragrance and quality for a longer time.

Extracting Essential Oil: If your goal is to extract essential oil out of lavender, make sure you remove the flower buds from the stems once you have completed dry process. Use the proper extraction techniques for obtaining top-quality lavender essential oils.

Section 5: Harvesting Considerations

Be aware of these additional factors in the process of harvesting lavender

Field Conditions: Don't harvest lavender in damp or wet weather conditions. The excessive moisture may impact the quality of lavender and can increase the chance for rot and mold.

Avoid overharvesting: Take care to not over-harvest lavender flowers, particularly in the first few years of their setting up. Let the plants grow and develop robust root systems prior to taking a long-term harvest.

Sustainable Harvest: Follow sustainable harvesting methods to guarantee longevity in the health and efficiency for your farm's lavender. Beware of excessive harvesting or non-sustainable techniques that could compromise the plant's ability to grow and regenerate.

Quality Control: Continually evaluate the condition of your freshly harvested lavender blooms. Remove any flower that shows indications of mold, discoloration or damage in order to keep high standards of quality.

Harvesting Record: Make precise records of harvesting practices, which include the dates, quantity as well as specific varieties of lavender that you have harvested. These records can assist you to monitor the effectiveness of various kinds of lavender and improve your harvesting methods for the future.

Chapter 14: Post-Harvest Handling: Drying And Storing Lavender

A proper post-harvest care is essential in preserving the value as well as the fragrance of lavender. Properly drying and storing lavender will ensure that it retains its beautiful color, strong aroma as well as the essential oil content. In this article we'll go over the fundamental steps to follow and recommended methods for drying and storage of the lavender following harvesting for your farm to grow lavender.

Section 1: Preparing Lavender for Drying

Prior to beginning drying, make sure you make sure that the lavender you have picked is prepared for the most optimal outcome. Follow these actions:

Harvesting Time: Harvest lavender at the time that flowers are at the prime stage of blooming. This guarantees the finest essential oil and scent. Check out Chapter 13 for detailed guidelines regarding the time and method for harvesting lavender.

Bundling: bundle the picked lavender stems into smaller bundles. Collect around 100 - 150 stems, aligning them with the flowers and then adjusting the length of stems to achieve consistency. Attach the bundle using twine or rubber bands with a trail for hanging.

Cleaning: Clear any particles that has damaged or discolored the flowers, or pests out of the lavender bouquets. Carefully shake or lightly scrub the flower spikes in order to loosen the loose materials.

Section 2: Drying Lavender

It is crucial to dry lavender correctly to keep its beautiful aroma, color, as well as its essential oil amount. Make sure to follow these guidelines for efficient drying of lavender:

Hang drying: hang the lavender bouquet upside down in a dark, cool well ventilated area. Pick a place that is protected from intense sunlight and temperature. The proper flow of air is essential to aid in drying, and to prevent the growth of mildew or mold.

Proper Ventilation: Be sure the area of drying has enough ventilation that allows for air flow around lavender bundles. This prevents the accumulation of moisture, and also guarantees the uniform drying.

Guard against Dust The best way to prevent the accumulation of dust on dried lavender flowers, you can cover them with a thin paper bag or cloth. This layer of protection allows an airflow that is appropriate while keeping the lavender fresh.

The time it takes to dry Lavender usually takes between approximately 1 to 3 weeks to completely dry according to the weather conditions. The flowers will be dried when the stems are fresh and crisp. The blooms retain their vivid scent and vibrant color.

Continuous Monitoring: Frequently check the lavender bouquets that are drying to see if there are any indications of mildew or mold or discoloration. If you find any concerns you should address them immediately through adjusting drying conditions or taking out the any affected bundles.

Section 3: Storing Dried Lavender

When the lavender bouquets have completely dried, it's important to properly store them so that they can maintain their high-quality scent, aroma, and oil composition. Follow these guidelines:

Destemming: During the drying process is complete, take the dried lavender blooms from the stems. It is done by firmly securing the stems while running your fingers down to remove the flower. Place the bouquets into a container that is clean, and then discard the stems.

Storage in airtight containers: Place the dried flowers of lavender in airtight containers to protect them from being exposed to light, water and air. Select glass or tin container with tightly-fitting lids. Do not use plastic containers since they can hold water and reduce the quality of dried lavender.

Storage that is cool and dark Containers of lavender that have dried in a dark, cool space, away from direct sunlight and heating sources.

Exposure to too much light and temperatures can cause the flower to fade, and eventually lose their aroma.

Labeling and date: Label each container with dried lavender that contains the lavender's specific variety as well as the date when it was harvested. This will allow you to keep track of the freshness and quality of the lavender you have purchased over time.

Use and Rotation: Make use of the most seasoned bunches of dried lavender the in the beginning to keep a cycle and guarantee the freshness. When stored properly, dried lavender will keep its scent and quality for 1 year or even longer.

Section 4: Additional Considerations

Take into consideration the following other aspects to dry and store lavender

Quality Control: Check regularly dried lavender flowers to determine if there are evidence of mold, discoloration or loss of smell. Removing any affected flower as quickly as you can in order to avoid spreading concerns.

Marketing and packaging: Take into consideration the presentation and packaging of your dried lavender, especially for commercial purposes. Make attractive and appealing packaging, with labels that emphasize the varieties of lavender, their aroma, as well as its potential applications. It will increase the appeal of the lavender items you sell.

Essential Oil Extract: If you are planning to extract essential oils from the lavender you have, be sure that dried flowers are stored properly so that they can preserve their essential oil contents. Use the appropriate extraction techniques for obtaining top-quality lavender essential oils.

Culinary Utilization: Correctly dried and stored lavender may be utilized for cooking. Make sure that the lavender isn't contaminated by chemicals or pesticides, and then use the lavender sparingly while baking and cooking to add the aromas and flavors.

Chapter 15: Identifying Value-Added Products Derived From Lavender

Agriculture Lavender farming does not just provide the opportunity to grow an exquisite and aromatic crop, but also offers access to a variety of value-added items. Through the transformation of lavender into a collection of exclusive and desirable products that you can expand your offerings of goods, boost profits, and increase the general attractiveness of your lavender farm. In this article we'll look into the world of high-value items derived from the cultivation of lavender.

Section 1: Lavender Essential Oil

Essential oil of lavender is one of the most well-known and diverse products that originate from lavender. It is renowned for its soothing and relaxing effects the essential oil of lavender is used in a variety of potential uses. Take a look at the following

Extraction Methods: The essential lavender oil is extracted using several methods, including steam distillation and solvent extraction.

Choose the one that is in line with your resources as well as quality standards.

Qualitative Control: Pay particular at the high-quality of the lavender flowers that are used in essential extraction of oils. Pick flowers with a higher essential oil content, and make sure to use that they are dried and stored properly to ensure their high-quality.

Application: The essential oil is used in a variety of products, such as candles, aromatherapy oils, soaps, lotions and even room sprays. Find different uses and develop distinctive blends that cater to the diverse preferences of customers.

Section 2: Lavender Culinary Products

The unique herbaceous and floral taste is a great match for cooking creations. When you incorporate lavender in edible items, you are able to offer your customers an experience that is unique to the culinary world. Take a look at the following examples:

Culinary Lavender: Use culinary-grade lavender flowers for creating edible products. Be sure to ensure that the lavender is been cultivated without the application of pesticides that harm the environment or chemical.

Culinary Lavender Recipes: Play around using recipes that are infused with lavender, like lavender-infused honey the lavender-scented shortbread cookie, olive oil that is lavender-infused and vinegar with lavender, as well as lavender-infused syrups. Create a set of recipes that demonstrate the many uses of lavender cooking as well as baking.

Identification and packaging. Pay particular attention to the labels and packaging of the culinary lavender items you purchase. It is important to clearly identify the type of lavender utilized, as well as the recommended culinary uses as well as any preparation or storage guidelines. Utilize attractive packaging to improve the appeal, and also marketability of your product.

Section 3: Lavender Bath and Body Products

The soothing aroma and calming qualities of lavender are a major ingredient in products for bath and body. When you create lavender-infused soaps and lotions, bath salts as well as other products it is possible to provide customers with the ultimate luxury and feeling. Take a look at the following examples:

Handmade Soaps: Make handcrafted soaps infused by essential oils of lavender and dried flowers of lavender. Explore different combinations and textures, to suit the different types of skin and preference.

Lavender Balms and Lotions Make moisturizing creams or body butters as well as balms that are infused with the essential oils of lavender. The soothing qualities of lavender make these items ideal for relaxing as well as relaxation routines.

Bath Salts and Bath Bombs Make lavender-infused bath salts and bath bombs providing customers with a spa-like luxury at your home. Explore various combinations of salts, oils as well as dried lavender petals to create relaxing bathing rituals.

Section 4: Lavender Home Fragrances

The refreshing and soothing scent of lavender is wanted for its ability to create an ambiance of peace in your home. Consider the design of scent-laden candle, sprays for rooms and diffusers that can enhance the scent of lavender in your living space. Think about the following aspects:

Lavender Candles: Create candles that smell of lavender using premium natural oils and natural waxes. Try different candle designs, sizes and options for packaging that appeal to a broad variety of consumers.

Sprays for rooms and diffusers Make lavender-infused sprays for rooms as well as diffusers for an inviting and relaxing ambience at home. Try different scents and designs for packaging to appeal to different tastes of your customers.

Gift Sets and Gift Sets and Home Fragrance Bundles Pack the lavender scent of your home with attractive gift packages or sets. You can combine lavender-scented candles or room sprays with diffusers to create a complete aromatherapy experience at home.

Section 5: Lavender Crafts and Decorative Items

Its beautiful and versatile nature makes it a great source of material for creating decorative objects and other unique products. Think about the following:

Lavender wreaths and bouquets Lavender wreaths and bouquets: Make gorgeous lavender bouquets and wreaths by using dried lavender flowers as well as complementary foliage. These items of decor are available as stand-alone items, or included in gift sets.

Lavender Sachets as well as Potpourri Make sachets and potpourri bags with lavender dried flowers making it possible for customers to experience the pleasant scent in their wardrobes as well as in drawers and vehicles. Explore different fabrics as well as decorative accessories to add an elegant touch to your items.

Chapter 16: Lavender Oil Extraction

Essential oil of lavender is a sought-after product that is made from the flowers of lavender. Due to its distinctive fragrance as well as numerous health benefits it is great demand within the areas of aromatherapy, personal care as well as the fragrance industry. In this section we'll explore various methods and tools that are used to extract lavender oil which allows you to create premium oils from your own lavender farms.

Section 1: Steam Distillation

Steam distillation is by far the most commonly used method for extraction of lavender essential oils. This method is a traditional one that ensures retention of lavender's essential substances and healing properties. Follow these steps to steam distillation

Equipment:

A. Distillation Equipment: Make use of an apparatus for steam distillation that consists of a boiler condenser, distillation chamber as well as a collector vessel. Make sure that the apparatus is made of food grade glass or stainless steel in order for purity, and to prevent chemical reactions.

b. Water Source: You must have that you have a water source reliable to generate steam for the process of distillation.

Preparation:

a. Harvesting: Harvest the lavender flowers when they are at their most blooming phase, to guarantee the most essential oil and scent.

B. Drying Lavender flowers are dried by using the correct drying methods discussed in chapter 14 to get rid of the excess moisture while ensuring efficient oil extraction.

Distillation Process:

a. Charge the Distillation Chamber Then, place the lavender flower arrangements that have dried in the distillation chamber on the device. Make sure that they are distributed evenly in order to allow for optimum steam flow.

b. Generating steam: Fill the steam boiler with hot water and then make it hot enough to create steam. The steam is then able to pass into the distillation chamber transporting the essential oil of the lavender flower.

C. Condensation and Collection The steam carrying the necessary oil will go through the condenser before it is condensed into liquid. The condensed water and oil mix is then collected into the separate vessel. Since the oil is floating in the liquid, it could be separated with the funnel separator or decantation technique.

D. Separation of oil and Water after the distillation process is completed, allow the oil and water mix to get settled. The oil that is essential will then rise onto the surface and make it simpler to separate and separate from water.

Storage:

a. Keep the essential oil of lavender in dark glass bottles or containers with amber colors to keep it safe from light and preserve its strength.

A. Label the container using the dates of extract along with the lavender type used to determine the quality and freshness of the oil.

Section 2: Solvent Extraction

Solvent extraction can be a viable technique used to perform lavender oil extraction, especially in cases of large-scale production. Solvent extraction involves the use of solvents to dissolve the essential oil of the flowers of lavender. Follow the next steps to extraction of solvents:

Equipment:

a. Extraction Unit: Make use of an extraction device that is specifically designed to extract solvents, and comprises a cylindrical vessel that houses a reservoir for solvents an condenser, as well as an extraction vessel. Make sure that the apparatus is constructed from materials that are compatible with essential oils as well as solvents.

b. Solvents: The most common solvents used for the extraction of lavender oils are the ethanol and hexane as well as supercritical carbon dioxide (CO2). Select the right solvent to the needs of your extraction, take into consideration their safety as well as their environmental impacts.

Preparation:

a. Harvesting: Harvest the lavender flowers in their prime bloom stage to ensure the best essential oil and scent.

B. Drying The lavender flowers should be dried by using the correct drying methods discussed in the chapter to get rid of excessive moisture. This will also facilitate extraction of oil.

Extraction Process:

A. In the process of loading the Vessel Lavender flowers are placed in the Vessel. flowers inside the extractor vessel of the device. Make sure that they are evenly distributed so that they are in contact with the solvent.

B. Extracting Solvents: Place the solvent you have chosen in the extractor vessel so that it comes into contact with the flowers of lavender. The solvent dissolves the oil essential components of the flower, creating an emulsion.

C. Separation and Recovery After an adequate extraction time then separate the solvent-oil mix from the discarded flowers. It can be accomplished by filtering or centrifugation techniques to eliminate the plants' material.

D. Solvent Recovery Remove essential oils from solvent with methods like evaporation or distillation in order to extract the essential oils of lavender.

Quality Control and Storage:

a. Conduct quality testing like gas chromatography in order to verify the quality and purity of the lavender essential oil.

b. Keep the essential oil of lavender in dark glass containers or amber colored bottles to shield it from sunlight and to preserve the potency.

C. Label the bottles in the order of date as well as the lavender type used to determine the quality and freshness of the oil.

Section 3: Cold Press Extraction (Enfleurage)

Cold-press extraction often referred to as enfleurage, the most traditional method for extraction of essential oils from delicate plants like lavender. Though it's not commonly utilized to extract lavender oils today but it's still a feasible option to produce small quantities or for artisanal purposes. Check out the steps to follow to extract cold presses:

Equipment:

a. Glass Plates Use glass plates with a layer of non-odour pure vegetable fat for example, coconut oil or clarified butter. The fat will soak up the essential oils from the flowers of lavender.

b. enclosed container: Put the lavender blooms between glass plates to form an arrangement. Place the flower arrangement within a sealed container to avoid the exposure to contaminants and air.

Extraction Process:

and. The loading of the stack Layer lavender flowers on the glass plate, then cover the top with a thin coating of fat. Repeat the process by switching between the layers of flowers and fat until the stack is completed.

B. Time to maturity The stack should sit for a few days or weeks in a dark, cool area. In this time the essential oil that comes from the flowers of lavender slowly diffuses through the fat.

C. Renewing the flowers Renew the flowers regularly with new ones for maximum absorption of oil.

D. Extract: Following the time of maturation, take the fat off of the heap, which includes the essential oil infused.

E. Recuperation of Oil: Remove the oil essential to the fat with techniques like steam distillation or solvent extraction.

Quality Control and Storage:

a. Conduct quality testing to confirm the quality and purity of the essential oils of lavender.

b. The essential oil of lavender should be stored in dark glass bottles or amber containers so that it is protected from the sun and preserve the potency.

C. Label the bottles in the order of date as well as lavender varieties used to monitor the quality and freshness that the oils are.

Section 4: Safety Considerations

If you are working with extraction techniques and equipment, it is essential to consider safety as a top priority. Take into consideration the following safety concerns:

Ventilation: Be sure to have adequate ventilation throughout the extraction zone in order to stop the formation of solvent vapors and steam.

Protective Gear: Always wear safety gear like goggles, gloves and the lab coat, while working with solvents or using extraction equipment.

Safety in Fire: Follow the procedures for protecting yourself from fire for working with solvents and heating sources. Be sure that the extraction zone has extinguishers for fire and adheres to local fire safety rules.

Storage and Handling: Keep essential oils and solvents inside containers with a valid seal, far from flames, heat and the direct light of sunlight. Be sure to follow safe techniques for handling in order to avoid spills or accidents.

Chapter 17: Crafting Lavender-Based Cosmetics And Skincare Products

Lavender is not just known because of its unique scent however, it also has numerous benefits to the skin. Incorporating lavender into products for skincare and cosmetics and products, you will be able to benefit from its calming, revitalizing and aromatic properties to produce an array of exquisite products for your skin. In this section we'll look at how to create cosmetics and skincare items that contain lavender that will allow you to tap the power of lavender for improving beauty and wellbeing.

Section 1: Lavender-infused Facial Cleansers and Toners

Toners and facial cleansers that are that contain lavender are an easy but effective method to clean, cleanse, and revitalize skin. Take note of the following:

Ingredients:

a. Lavender Hydrosol is a byproduct of the steam distillation process is a soft and soothing ingredient that soothes and moisturizes the skin.

b. Essential Oil of Lavender: The use of essential oils like lavender offers antibacterial and soothing properties that help to maintain a clean and healthy complexion.

C. Cleaning Agents Pick pure and gentle cleanser agents such as coconut-based soapwort surfactants, or gentle soapwort extracts that remove dirt without leaving behind its natural moisture.

Formulation:

a. Blend lavender hydrosol with lavender essential oil, and other cleansing agents with the right amounts to produce a mild and efficient facial cleanser, or toner.

b. You can add other ingredients that are skin-loving for example, aloe gel, chamomile extract or witch hazel to increase the cooling and clarifying benefits.

C. Pack the cleanser and toner in appropriate containers, particularly dark glass bottles to keep the components safe from the sun and preserve their efficiency.

Section 2: Lavender-infused Facial Masks and Scrubs

Masks and scrubs for the face made with lavender provide an aromatic and rejuvenating sensation while also promoting a healthier and beautiful skin. Take note of the following:

Ingredients:

a. dried lavender flowers: Add small pieces of crushed or finely ground lavender blooms to provide soft exfoliation, and to enhance the aroma.

b. lavender essential oil: addition lavender essential oil improves the skin soothing and relaxing effects of the scrubs and masks.

C. Organic Exfoliants Pick mild natural exfoliants like finely crushed Oats, almond meal or jojoba beads. These are used to eliminate dead skin cells to show a more supple skin.

Formulation:

a. Mix finely crushed or dried flower petals of lavender essential oils of lavender and natural exfoliants, with suitable base ingredients, like yogurt, clay honey, honey or aloe vera gel to make facial masks or scrubs.

b. You can consider adding other skin-nourishing ingredients such as kaolin clay, green tea extract and vitamin E oils, in order to increase the effectiveness of your scrubs and masks' beneficial effects.

C. Place the masks or scrubs into hygienic and safe containers with clearly marked uses and frequency of use.

Section 3: Lavender-infused Moisturizers and Serums

Creams, serums, and moisturizers with lavender provide moisture, nourishment as well as anti-aging benefits for the skin. Take note of the following:

Ingredients:

a. essential oil of lavender: addition lavender essential oil gives it softening, moisturizing and regenerative qualities to formulations.

b. Carrier oils: Pick the most nourishing, skin-friendly and skin-friendly oils such as rosehip seed oil, and argan oil to provide moisture and vital oils for skin.

C. Vitamin C: Think about adding ingredients such as hyaluronic acids or vitamin C Peptides to increase collagen production, boost skin elasticity and decrease wrinkles.

Formulation:

a. Blend lavender essential oil with carrier oils and various substances for skin rejuvenation in the right proportions to make moisturizers or serums.

It is b. Modify the formula to address specific skin issues for example dryness, ageing or sensitivity through adjusting the quantity of active ingredients or adding other botanical extracts.

C. Place the moisturizers and serums in airtight containers typically with an pump or dropper to ensure simplicity of use as well as for keeping the freshness of your product.

Section 4: Lavender-infused Lip Balms and Body Lotions

The lavender-infused lip balms and body lotions provide relief and relaxation, and keep skin moisturized and safe. Think about the following

Ingredients:

a. Lavender Essential Oil: use of essential oils of lavender imparts its calming and moisturizing qualities that leave the skin and lips smooth and soft.

b. Natural Emollients: Choose natural emollients like shea butter or cocoa butter or beeswax. These ingredients provide lasting moisture and protect for your lips and body.

C. skin-soothing extracts: You might want to consider adding ingredients such as calendula or chamomile extract aloe vera gel, to soften and regenerate the skin.

Formulation:

a. Mix natural emollients like beeswax and shea butter together with the nourishing oils such as coconut oil and almond oil.

b. Incorporate lavender essential oil and moisturizing extracts for the skin to the mixture. Stir thoroughly to ensure an even dispersion of the ingredients.

C. The mixture should be poured into the appropriate lipstick tubes or containers, and let it set prior to sealing.

Section 5: Lavender-infused Fragrances and Perfumes

Lavender fragrances and scents provide the most relaxing and relaxing sensation. Take a look at the following

Ingredients:

a. essential oil of lavender: addition of essential oils of lavender serve as the principal aromatic ingredient that gives the typical and relaxing lavender aroma.

B. Essential oils that complement each other: Choose other essential oils like bergamot, geranium or rosemary to make unique and well-balanced scent profiles that compliment the lavender scent.

C. Alcohol, or Carrier Oil: Choose an suitable base for your fragrance for the fragrance, like perfumer's Alcohol or skin-compatible carrier oils depending on the purpose of roll-on or spray applications.

Formulation:

a. Mix lavender essential oil and the other essential oils according to the ratio desired, in order for a balanced scent blend.

b. Mix the scent using perfumer's alcohol or carrier oil and adjust the amount according to your the individual's preference.

C. Keep the scent in glass containers that are dark and safe which will allow it to develop for several weeks in order in order to develop its aroma to the maximum.

Chapter 18: Lavender In Culinary Delights

The lavender plant isn't just adored by its aromatic blooms or essential oil, but also due to its broad application in the kitchen. The addition of a hint of lavender into your cooking improve the taste, add an appealing aroma as well as add an interesting taste to many food items and drinks. In this section we'll look into the captivating aroma of lavender when it comes to food preparation, providing the recipes and suggestions to incorporate lavender into your cooking and please the palates of the people who eat the creations you make.

Section 1: Lavender-infused Beverages

Lavender is a great drink to add giving a flowery note and an element of calm to the drinks you choose. Check out these recipes

Lavender Lemonade:

Ingredients:

1 cup fresh squeezed lemon juice

4 cups of water

1/2 cup of lavender syrup (made through boiling water, sugar and dried flowers of lavender)

Ice cubes

Instructions:

In a pitcher, mix the juice of a lemon as well as the lavender syrup.

Mix well and thoroughly to combine all the ingredients thoroughly.

Serve on ice, And garnish it with a fresh sprig of lavender, or a slice of lemon.

Lavender Iced Tea:

Ingredients:

4 cups of water

4 lavender tea bags, or 2 tablespoons of dried lavender flowers

Honey, sugar or both to suit your tastes

Fresh lavender sprigs for garnish

Instructions:

The water should be brought to a simmer and then put in the lavender tea bags or dried lavender blooms.

Keep stirring for 5-10 minutes or until the strength you desire is reached.

Take the tea bags out and take the lavender flowers out of them.

Add honey or sugar depending on your taste.

The tea should cool, then refrigerate it until it is chilled.

Serve on ice with decorate with lavender leaves.

Section 2: Lavender-infused Baked Goods

Lavender is a wonderful ingredient to add a subtle scent and floral taste to various baked products, from cakes and cookies, to pastries and bread. Take a look at these recipes:

Lavender Shortbread Cookies:

Ingredients:

1 cup of unsalted butter softened

1 Cup powdered sugar

2 cups all-purpose flour

2 tablespoons dried lavender flowers

Instructions:

Preheat the oven to 350 degF (175degC).

A mixing bowl combine the butter and powdered sugar, until light and airy.

Add the flour, as well as dried lavender flowers until blended.

The dough is then rolled into smaller balls, then place them lightly on baking sheets covered by parchment.

Bake for 12-15 minutes or until edges have been lightly caramelized.

Cool the cookies by placing them on wire racks prior to serving.

Lavender Honey Bread:

Ingredients:

3 cups all-purpose flour

1 tablespoon dried lavender flowers

1 teaspoon salt

2 teaspoons of active dry yeast

1 1/4 cups of warm water 1 1/4 cups warm

2 tablespoons of honey

Two tablespoons of olive oil

Instructions:

In a mixing bowl large enough put the flour in a large mixing bowl, along with dried lavender flowers, as well as salt.

In a small bowl, mix the yeast in hot water. set it aside for 5 minutes, until it is it is frothy.

Include the honey as well as olive oil in the yeast mix and mix thoroughly.

Then, gradually add wet ingredients to dry ingredients and mix until dough begins to form.

The dough should be kneaded on a surface that is floured for 5 to 7 minutes or until the dough is elastic and smooth.

The dough should be placed in an oil-sprayed bowl, then cover with an unclean kitchen towel and allow it to rise in a warm area for approximately 1 hour or until the dough has grown to a size of.

Preheat the oven to 375 degF (190degC).

Make sure to punch the dough down and make it into an oval loaf. The loaf should be placed in an oven-proof loaf pan that has been greased.

The pan is covered with an old kitchen towel, and allow the dough rest for another 30 minutes.

Bake for 30 - 35 minutes and until bread has turned golden brown and it sounds hollow by tapping it on the top.

Let the bread sit on the wire rack to cool prior to slicing.

Section 3: Savory Lavender-inspired Dishes

Lavender is also a great ingredient to incorporate into dishes that are savory, bringing an unique taste and providing a hint of sophistication in your food creations. Take a look at these recipes:

Lavender Roasted Chicken:

Ingredients:

1 Whole chicken

2 tablespoons dried lavender flowers

4 cloves garlic, minced

Two tablespoons of olive oil

Salt and pepper as desired

Instructions:

Preheat your oven to 425 deg F (220degC).

Rinse and clean it with the help of paper towels.

In a small bowl combine the dried lavender blooms, minced garlic, olive oil salt and pepper.

Massage the lavender paste on the chicken. Be sure you coat it in a uniform manner.

Set the chicken in an oven tray and cook for approximately one hour or until temperature inside reaches 165degF (74degC).

Take it out of the oven, and let it be for about a minute prior to cutting it into pieces.

Lavender-infused Salad Dressing:

Ingredients:

1/4 cup extra virgin olive oil

2 tablespoons of white wine vinegar

1 teaspoon Dijon mustard

1 teaspoon honey

1 teaspoon of dried lavender flowers

Salt and pepper as desired

Instructions:

In a small dish, combine the olive oil white wine vinegar Dijon mustard, honey the dried flowers of lavender, salt and pepper until thoroughly mixed.

Check the seasonings and taste, then adjust them according to your preference.

Serve the dressing with a hint of lavender on your salad greens and vegetables.

Section 4: Lavender-infused Condiments and Spreads

Lavender adds a delicate and fragrant twist to various sauces and condiments, increasing their flavor, and delivering delight to your palate. Take a look at these recipes:

Lavender-infused Honey:

Ingredients:

1 cup honey

2 tablespoons dried lavender flowers

Instructions:

In a clean jar blend the honey and the dried flowers of lavender.

Cover the jar with a lid and let it sit for at minimum one week in a dark, cool area.

Take the lavender flowers out and place the lavender-infused honey to a clean container.

Utilize the lavender-infused honey an ingredient in your tea as a sweetener or drizzle it over desserts or used as a topping to yogurt.

Lavender-infused Butter:

Ingredients:

1 cup unsalted butter softened

2 tablespoons dried lavender flowers

1 tablespoon honey (optional)

A pinch of salt

Instructions:

A mixing bowl mix together softened butter and dried lavender flowers honey (if wanted) as well as the pinch of salt to make sure it is all ingredients are well incorporated.

The lavender-infused butter can be placed on a sheet of parchment paper or wrap it in plastic.

Chapter 19: Examining The Medicinal Properties Of Lavender

Known for its beautiful scent and bright lavender a bloom is a plant that has been cherished throughout history for its healing qualities. Apart from its attractive appearance it also has a vast variety of benefits for medicinal use which have been acclaimed and used in a variety of conventional healing methods. In this section we'll explore the therapeutic properties of lavender and explore its use for aromatherapy, herbal medicines as well as personal care.

Section 1: Aromatherapy and Stress Relief

Lavender is known for its relaxing and stress relieving qualities, which makes it an extremely popular option for aromatherapy. Think about the following aspects:

Relaxation and Sleep Aid:

Essential oils of lavender have an aroma of calm that encourages calm and promotes sleep. The sedative qualities of lavender can reduce insomnia and increase sleep quality. sleep.

Incorporate a few drops lavender essential oil into the diffuser, or sprinkle it on your mattress to provide a serene atmosphere which is conducive for a peaceful sleep.

Anxiety and Stress Reduction:

The aroma of lavender can assist in decreasing stress and anxiety levels. Lavender's scent is proven to have an effect of calming in the brain. It can also promote an overall sense of peace and happiness.

Infuse a room spray with lavender or mix essential oils of lavender with carrier oils to make an energizing massage oil.

Section 2: Skin Care and Wound Healing

Lavender is a potent skin healer with a variety of properties that make it a useful ingredient in skin care products as well as treatment for wounds. Take a look at the following benefits:

Anti-inflammatory and Antiseptic Properties:

Lavender essential oil is antiseptic and anti-inflammatory properties. aid in healing and soothing many skin ailments, like acne, eczema, as well as minor burns.

Mix a few drops lavender essential oil into a carrier oil before applying the oil topically on areas of concern for reducing inflammation and to promote healing.

Skin Regeneration and Scar Reduction:

The regenerative power of lavender is evident It can help in the reduction of marks and promote healthy skin.

You can make a lavender-infused lotion cream. Apply it on parts of the skin that are damaged for a better healing process and to improve the overall appearance of skin.

Section 3: Headache and Migraine Relief

Lavender has been utilized since the beginning of time as a natural cure for migraines and headaches. Take a look at the following benefits:

Analgesic Properties:

The analgesic qualities of lavender will help ease migraines and headaches. Application of the oil through inhalation or on the skin lavender essential oil could provide relaxation from stress as well as help to relax.

Inhale lavender essential oil and apply diluted solutions to foreheads, temples, and the back of your head for relief from headache symptoms.

Stress and Tension Reduction:

The soothing scent of lavender can to reduce tension and stress, which can trigger migraines and headaches.

Try relaxing techniques with lavender infusion like lavender-scented baths or massages to ease the symptoms of headaches.

Section 4: Digestive Support and Relaxation

Lavender is a popular herb for its ability to aid digestion as well as aid relaxation. Think about the following benefits:

Digestive Aid:

Lavender may help to soothe the digestive tract, and reduce symptoms like bloating, indigestion and gas.

Enjoy a cup of tea with lavender infusion after meals or mix lavender essential oil and an oil carrier and rub it on the abdomen to ease digestion.

Relaxation and Stress Reduction:

The calming effects of lavender extend into the digestive tract and help ease digestion issues caused by stress.

Integrate lavender aromatherapy and lavender-infused products for personal care in your daily self-care routine for relaxation and improve digestion.

Section 5: Pain Relief and Muscle Relaxation

Lavender has been utilized since the beginning of time to ease tension and improve muscle relaxation. Take a look at the following

Muscle Relaxation:

Essential oil of lavender is known to ease muscle tension, relax muscles tension and relieve muscular pains and aches.

Incorporate a couple of drops of essential lavender oil into your warm bath, or mix it in with an oil carrier for relaxing massage.

Joint Pain Relief:

Lavender's anti-inflammatory properties may help decrease joint inflammation as well as ease joint pain that comes with arthritis, or stiff joints.

Massage a lavender-infused balm cream onto the region to ease the pain.

Section 6: Respiratory Health and Congestion Relief

The beneficial effects of lavender are seen in the respiratory system and may help alleviate congestion. Think about the following benefits:

Respiratory Support:

The aroma of lavender essential oil may aid in relieving respiratory ailments like colds, coughs and congestion in the sinuses.

Inhale lavender essential oil and to create steam by using a few drops lavender essential oil into an ice-cold bowl while inhaling steam.

Allergy Relief:

The anti-inflammatory properties in lavender may help to reduce symptoms associated with allergic reactions. They also provide relief from itchy and nasal congestion. eyes.

Utilize lavender-scented items in your residence, including lavender sprays, or air fresheners for a non-allergenic and tranquil space.

Chapter 20: Creating Lavender Sachets And Potpourri

Potpourri and lavender sachets can be a wonderful way to appreciate the wonderful scent and soothing qualities of lavender. They can be utilized to freshen rooms and wardrobes or even drawers, as well as giving a touch of class to the decor of your home. In this article we'll look at how to make lavender potpourri and sachets which will allow you to fill your home with the scent and beauty of lavender.

Section 1: Lavender Sachets

Lavender sachets can be described as small, woven bags that are filled with dried lavender flowers which release pleasant scents when they are squeezed lightly or placed inside enclosed spaces. Take note of the steps needed to make lavender sachets

Material Selection:

Pick a material that can be used for sachets such as light cotton, muslin or linen. Choose fabrics with a variety of shades or patterns for an some flair.

Cutting and Sewing:

Make the material into rectangular or square pieces. Allow enough room to sew and fill in the spaces.

Put two pieces of fabric together and place the pattern or coloured sides facing one opposite.

Make three seams on the fabric together, leaving one side free to fill with.

Filling the tank with Lavender:

The sachet should be filled with dried lavender flowers. Make sure they're evenly distributed.

In order to make filling more easy make a funnel with the paper you have or a funnel-making tool.

Be careful not to overfill so that the sachet will be able to be squeezed while maintaining its form.

Closing the Sachet:

Fold the exposed side of the fabric towards you and then stitch it to close with a thread and needle or sewing machine.

Attach a ribbon of a smaller size or a string around the sachet, to give it an attractive accent.

Usage:

The lavender sachets can be placed in cupboards, drawers or in storage spaces to give an inviting scent as well as deter insects.

It is also possible to put sachets beneath your bed for peaceful sleep or keep them around in your backpack to enjoy a fresh scent during the day.

Section 2: Lavender Potpourri

Lavender potpourri is an enchanting mix of dried flowers, essential oils and herbs which releases an fragrance when it is displayed in the form of a sachet or bowl. Take note of the steps below for making lavender potpourri

Drying Lavender Flowers:

Lavender flowers harvested from the harvest after they have fully blossomed and in their best scent.

The lavender stems should be bundled in a bundle and then place them upside-down in a dark, cool and ventilated area. dry.

The flowers should dry completely, until they fall apart when pressed.

Gathering Other Ingredients:

Pick the right dried flowers, plants, and herbs to increase the scent and appeal of the potpourri. Some examples include rose petals the chamomile flower, rosemary as well as dry citrus peels.

Blending the Potpourri:

In a bowl large enough, blend the dried flowers of lavender along with the selected botanicals altering the ratios according to your personal preferences.

Include a few drops lavender essential oil to increase the aroma. Mix the ingredients gently in order so that essential oils.

Scent Preservation:

Keep the potpourri mix within an airtight container several weeks in order for the aroma to develop and increase in intensity.

It is important to shake the container regularly in order to distribute the essential oil and to encourage the mixing of aromas.

Display and Refreshment:

Put the potpourri in attractive containers, such as vases or bowls. You can also use sachets, and place them that you'd like to relax and enjoy the aroma.

To enhance the scent of the potpourri Add a couple of drops of lavender essential oil. Or lightly crush dried flowers to let their fragrance out.

Section 3: Customization and Variations

Making lavender sachets or potpourri is a way to create endless customization and variants. Use these suggestions for personalizing your creations

Mixing Fragrances:

Mix lavender with other complimentary scents such as vanilla, rose or citrus to make distinctive fragrance mixes.

Adding Botanicals:

Explore various dried herbs, flowers and spices for an element of interest to the visual and aroma of the delicious potpourri.

Enhancing Aesthetics:

Decorate sachets with buttons, ribbons, or lace to give them the perfect touch of class.

Utilize colored or patterned fabrics to design visually appealing Sachets.

Gift Giving:

Potpourri and lavender sachets make beautiful gifts for family and family members. Pack them into attractive containers or bags, and add a personal notes.

Chapter 21: Lavender Farming For Essential Oil Production

Lavender cultivation for essential oil production can be an exciting and profitable venture. Essential oils of lavender are highly appreciated for its beautiful scent as well as its therapeutic qualities and its use in a range of fields such as the fields of aromatherapy, personal care as well as natural medicines. In this section we'll explore the fundamental steps and factors that are involved in the cultivation of lavender specifically designed for the production of essential oils and help you embark in a profitable journey to become a producer of lavender oil.

Section 1: Lavender Varieties for Essential Oil Production

The selection of the appropriate lavender species is vital to ensure a successful vital oil creation. Be aware of the following points when choosing lavender varieties

High Essential Oil Content:

Find lavender varieties that are known for their essential oil content, for example Lavandula angustifolia (True Lavender) or Lavandula intermedia (Lavandin).

Lavandula angustifolia species including "Munstead" and "Hidcote," have become famous for their excellent essential oils of high quality.

Climate and Adaptability:

Select lavender varieties that are suitable for your environment and climate.

Be aware of factors such as temperature tolerance, cold hardiness and the resistance to diseases and pests.

Yield and Harvest Timing:

Choose the varieties of lavender which provide an adequate amount of fragrant flowers that contain essential oils.

Choose varieties that have a staggered flower time to guarantee a long period of harvest.

Section 2: Lavender Farm Planning and Preparation

A well-planned and organized plan is crucial to a successful lavender farm focusing on oil production. Follow these actions:

Farm Site Selection:

Find a spot that has plenty of sunlight and well-drained soil because lavender flourishes in the sun. It does not like wet or muddy conditions.

Test soil samples to evaluate the soil's pH, as well as its the levels of nutrients, with a goal to achieve slightly alkaline soil pH between 6.5 to 7.5.

Field Layout and Plant Spacing:

Lay out your lavender field with care, taking into account aspects like accessibility to irrigation, ease of the management of weeds, as well as machinery accessibility.

Give enough distance between rows of lavender and the plants so that there is the circulation of air and for proper development. In general, rows should be separated by 3-4 feet apart. Plants should be separated by between 2 and 3 feet in the rows.

Soil Preparation:

Clean the soil by getting rid of dirt and weeds and tilling the soil in order to break down compacted soil. Add organic matter and compost in order for improved soil structure and fertility.

Resolve any soil issues discovered in soil tests with the appropriate amendment to soil like organic fertilizers or lime.

Section 3: Lavender Cultivation and Management

Achieving success in lavender cultivation and management is essential to maximize the production of essential oils. Take note of the following important elements:

Planting Lavender:

Cuts and seedlings of lavender in a well-groomed soil. Ensure that they're planted at the same level that their nursery pots.

The newly planted lavender should be watered well to set the soil around the roots to promote growth.

Irrigation:

Lavender needs moderate watering in particular during the growth stage.

Use a drip or micro-irrigation method to provide specific and reliable irrigation, since lavender is sensitive to excess water.

Weed Control:

Take effective measures to manage weeds including mulching, hand-weeding or even mechanical cultivation.

Use organic herbicides or approved ones when needed, observing the guidelines and limitations.

Pruning and Shaping:

Pruning lavender plants every year will preserve their form, stimulate the growth of more vigorous plants, as well as preventing dead, swollen stems.

Cut back in springtime or shortly after the first blooming flush. Remove one-third to half of the height.

Nutrient Management:

The majority of lavenders prefer soil that is in good conditions, so it doesn't need excessive fertilization.

Apply compost or organic fertilizers in a controlled manner, as necessary, based on soil testing results and nutritional needs.

Section 4: Lavender Harvesting and Essential Oil Extraction

The methods and timings for lavender harvesting as well as extraction of essential oils significantly affect the quantity and quality of oil that is extracted. Be aware of these actions:

Harvesting:

The lavender blooms of the harvest are at their peak Just before they begin to lose their color.

Utilize sharp pruning shears, or a sickle to trim the stalks of flower and leave a little of the stem tied to the flower.

Essential Oil Extraction Methods:

Pick a suitable essential oils extraction process, for instance steam distillation, or solvent extraction dependent on the amount of production as well as the resources available.

Steam distillation is by far the most popular method used for lavender oil extraction which uses steam to extract the oil from plant substance.

Oil Yield and Quality:

The quality and yield of the oil will vary based on various variables like the variety of lavender season, harvesting timing and methods of distillation.

Make sure that the correct distillation equipment, techniques and processes for post-distillation are used for obtaining top-quality lavender essential oil.

Section 5: Storage and Marketing

A well-organized storage system and efficient methods of marketing are crucial to keeping the high quality of lavender essential oil as well as achieving the market you want to reach. Be aware of these aspects:

Storage Conditions:

Keep lavender essential oil stored in dark glass bottles or containers with amber colors to keep it safe from heat and light and preserve its effectiveness.

Keep the oil stored in a dry, cool space far from direct sunlight as well as changes in temperatures.

Labeling and Packaging:

Correctly label your containers with the necessary information such as the name of the plant as well as the harvest year, lavender type, and extraction methods employed.

Think about attractive packaging and clear labels to attract your customers, and highlight the distinctive attributes of your essential lavender oil.

Marketing and Distribution:

Create a plan for marketing lavender essential oils, targeting prospective customers in areas like personal care, aromatherapy and even natural health.

Make use of websites, farmer's market, trade fairs, and local retail stores to show off and sell your lavender essential oils.

Develop relationships with prospective purchasers and establish a trustworthy company by highlighting the high-quality as well as the pure and beneficial properties of lavender essential oil.

Chapter 22: Marketing And Selling Your Lavender Products

After you've put in laborious work in cultivating lavender, and resulting in a wide range of wonderful lavender-based items after which it's time to concentrate on marketing and selling your lavender-related products. A well-planned marketing strategy along with a well-planned sales plan will help you reach your intended audience, creating the brand's reputation, and driving sales. In this section we'll look at the most important elements of selling and marketing lavender products. We will provide the essential information and techniques to make a mark on the marketplace.

Section 1: Branding and Product Differentiation

The creation of a brand that is strong and distinguishing your lavender-based product from the competition is crucial to entice buyers. Take note of the following points:

Define Your Brand Identity:

Find the distinctive characteristics and qualities that make the lavender products you sell distinct. Think about factors like the quality of your product, its sustainability, as well as workmanship.

Create a memorable brand narrative that highlights your love for the lavender industry and your dedication to high-quality as well as the value of your product.

Develop a Distinctive Brand Name and Logo:

Pick a memorable, meaningful branding name that connects with the people you want to reach.

Design a logo that is visually attractive that captures the essence of lavender. It also serves as a symbol of your brand's branding.

Product Packaging and Presentation:

Choose eco-friendly, attractive packaging that is in line with your image as a brand and boosts the perception of value for the lavender products you sell.

You can incorporate elements like gorgeous typography, beautiful imagery as well as sustainable products to make an pleasing and homogenous product line.

Section 2: Target Market and Customer Research

The understanding of your customer's needs and their needs is vital to a successful sales and marketing. Follow these steps:

Identify Your Target Market:

Discover the demographics, interest and buying habits of your potential customers.

Think about market segments like those who are interested in aromatherapy, eco-conscious buyers and those who are looking for the most natural and organic items.

Conduct Market Research:

Learn about customer trends as well as market trends and competition.

Make use of surveys, interviews as well as online research, to better understand the needs of consumers, their expectations, as well as price sensitivity.

Build Customer Relationships:

Get involved with your targeted customers via the use of social media, newsletters for email and blog articles that are informative.

Encourage testimonials and feedback to establish trust and build credibility.

Section 3: Online Presence and E-commerce

An online presence that is strong and well-established is crucial in this digital world. Use these tips for maximizing your web presence:

Build a Professional Website:

Develop a stunning and user-friendly site that highlights your lavender products, the brand stories, as well as purchasing choices.

Make sure your site is mobile-friendly optimised for SEO in order to enhance your website's its visibility.

E-commerce Platform:

Create an online storefront for online sales.

Clear product descriptions along with quality images and safe payment options that instil confidence in customers who purchase online.

Social Media Marketing:

Make use of social media channels like Instagram, Facebook, and Pinterest for showcasing your lavender items, post captivating content, and connect with your customers.

Make use of targeted advertising as well as influencer-related collaborations to broaden your reach and draw new clients.

Section 4: Retail Partnerships and Local Markets

Establishing relationships with retail stores as well as participating in local market events could increase your product's visibility and the sales. Take note of the following tactics:

Identify Retail Opportunities:

Find local shops, boutiques, gift stores spas, market stalls that are aligned with the market you want to target.

Reach out to potential partners for retail to present your items professionally. Highlight the unique selling points.

Wholesale Pricing and Terms:

Find wholesale prices that are competitive as well as attractive terms to attract retail stores to stock your products made of lavender.

Think about offering promotions or discounts on bulk purchases.

Participate in Local Markets and Events:

Make a display at farmer's markets and craft fairs as well as agricultural fairs to display and market your lavender-related products directly to your customers.

Interact with people, provide product samples, while sharing your expertise and enthusiasm for lavender cultivation.

Section 5: Collaborations and Partnerships

Collaboration with businesses that complement yours as well as forming strategic alliances can increase your client base and increase brand recognition. Take a look at the following strategies:

Collaboration with local artists:

Collaborate with local artisans for example, candle makers, soap makers or herbalists to develop products that are collaborative with lavender.

Promote and market the unique partnership, tap the customer base of each other.

Strategic Partnerships:

Look into partnerships with companies that are in similar industries like wellness centers, spas or even hotels, and offer lavender products as part of their service or other amenities.

Promote each other's product as well as share marketing efforts so that you each other benefit from greater publicity.

Influencer and Blogger Engagement:

Look for influencers or bloggers that have an interest in natural products, wellness or the sustainable lifestyle.

Partner with them in a collaborative way to highlight and discuss your lavender products. You can leverage their authority and reach to increase your reach.

Section 6: Customer Experience and Feedback

Offering a pleasant customer experience and taking advantage of customer feedback is essential to build loyalty as well as creating repeat customers. Think about the following methods:

Customer Service Excellence:

respond quickly and professionally promptly to any customer concerns, questions and comments.

Try to meet or exceed the expectations of your customers by providing personalized customer service, speedy shipment of orders and attention to the smallest of details.

Chapter 23: The Fascinating History Of Lavender

Lavender is an aromatic blooming plant which has charmed humans for centuries. Due to its stunning purple color and distinct aroma, lavender is used in medicinal as well as culinary uses for thousands of years. In this article we'll look at the fascinating background of lavender through its earliest beginnings and current uses.

Origins and Early Uses

The plant is believed to have come from the Mediterranean region, particularly in those areas that are now called Spain, Italy, and Greece. It was highly regarded by the early Greeks as well as Romans as they used it to serve a myriad of functions. It was used to fragrance bathwater as well as to freshen up the air of the homes as well as public areas. The Romans were also using lavender in cooking as well as for medical uses. It was believed that the plant could be a healing herb and could be used for treating various illnesses, ranging from headaches to stomach upsets, to burns, insect bites, and other injuries.

In during the Middle Ages, lavender became an ingredient used in perfumes and cosmetics. The plant was coveted due to its sweet aroma and also used to disguise the unpleasant smells. The plant was also employed during religious celebrations and believed to possess the power of spirituality. Some cultures believed that lavender was believed to keep away evil spirits as well as to guard against diseases.

Lavender in Modern Times

Even today, lavender remains extremely valued due to its numerous benefits. It is grown across the globe but France being the biggest producer. Lavender can be found in many different items, ranging including essential oils and fragrances to candles and soaps. It is also utilized to cook and bake especially for Mediterranean food preparation. Recently the use of lavender has grown to become the most popular ingredient in drinks like cocktails, and even other beverages.

One of the primary uses for lavender in the present is for aromatherapy. The herb is thought to provide a relaxing effect to the mind and body. It can help treat anxiety, stress, and insomnia. The essential oil of lavender is commonly utilized in massages and additionally used in the scenting of candles and bath products.

It is also utilized for beauty purposes. It is believed to possess anti-inflammatory properties, and it can be used for treating a variety of skin problems, including acne to Eczema. Essential oils of lavender are often utilized in products for skincare as it is believed to ease and soothe the skin.

In the end, lavender is one of the most popular plants among gardeners. It's easy to care for and requires minimal care, making it an ideal alternative for gardeners who are new to gardening. It is also appealing to pollinators like butterflies and bees, which makes it an excellent element to the garden.

Chapter 24: Why Lavender Is A Profitable Crop

Lavender is not just an attractive and aromatic plant, it could also turn into a lucrative product for entrepreneurs and farmers. Recently the market of lavender-based products has risen substantially, creating opportunities for those looking to cultivate lavender in order to make money. In this article we'll look at the potential market and trends for those who grow lavender.

Market Trends

The market for lavender-related products is on the rise in recent times, driven in part by the rising interest in organic and natural items. The plant is versatile which can be utilized in many different products including essential oils and fragrances to candles and soaps. Lavender can also be used for baking and cooking, specifically for Mediterranean food preparation.

One of the major factors driving the demand for lavender is the rising popularity of aromatherapy. The lavender plant is believed to possess a relaxing effect on body and the mind.

It is utilized to combat anxiety, stress, and insomnia. Lavender essential oil is commonly utilized in massages and often used to fragrance candles and bath items.

Alongside the health and beauty industry in addition to the beauty and wellness industries, lavender can also be found for food and beverages business. It is an ingredient that is widely used in drinks and cocktails It is also utilized in baking goods to add flavor and aroma different dishes.

Opportunities for Lavender Growers

A growing demand for lavender-based products opens up new possibilities for growers of lavender. There are a variety of options to earn money from the cultivation of lavender. Some of them include selling dried or fresh flowers, generating essential oils, or making value-added items like candles, soaps, and other beauty products.

Selling Fresh or Dried Flowers

One of the most straightforward methods to earn money from the cultivation of lavender is to market freshly cut or dried flower arrangements. The plant is popular to create flower arrangements. It could be offered to florists directly or to customers. Lavender flowers dried can be utilized in sachets, or potpourri. It can also be purchased at farmer's market as well as craft fairs and on the internet.

Producing Essential Oils

A second lucrative option for growers of lavender is to make essential oils. Lavender essential oil is among of the most well-known and extensively used essential oils for aromatherapy. Essential oils are created through distillation of the flowers from the lavender plant. They are sold to producers of products for aromatherapy or directly to customers.

Creating Value-Added Products

In addition, the lavender farmers can make value-added products with lavender such as candles, soaps and other beauty items. They can sell these products through online stores, craft fairs, or even at market stalls at farmers' markets. Producing value-added items allows farmers to boost the worth of their lavender harvest and to create an unique name that differentiates them from other cultivators.

Chapter 25: Choosing The Right Variety

The selection of the appropriate lavender cultivar is crucial to the successful operation of any business that is growing lavender. The different varieties of lavender have distinctive features that affect their yield, growth and even quality. In this article we'll discuss the various varieties of lavender, and the best way to select the best one to meet your requirements.

English Lavender (Lavandula angustifolia)

English lavender, which is also referred to as real lavender is the most commonly used cultivar of lavender used to be used commercially. It's indigenous to the Mediterranean region, and is coveted by its oil content as well as its sweet scent. English lavender can be described as a tiny shrub that usually will grow up to 3 to 4 feet.

English lavender is renowned for its distinctive scent and can be found in a range of items, such as essential oils and soaps, perfumes, and candles. It's also a favorite ingredient for cooking, and can be employed to flavour food items like tea, ice cream and Scones.

French Lavender (Lavandula dentata)

French lavender, also referred to as fringed, is indigenous to the Mediterranean region. It's distinguished by its serrated leaves as well as the purple flowers. It's a higher-growing variety of lavender. It typically grows at a height of 2-4 feet.

French lavender is famous for its distinctive scent. It can be found in a range of goods, such as essential oils, perfumes and bathing products. Also, it is a popular ingredient in cooking and employed to flavour meals like seafood, chicken, and even lamb.

Spanish Lavender (Lavandula stoechas)

Spanish lavender, sometimes referred to as butterflies lavender, also known as rabbit ears, is indigenous to the Mediterranean area and is distinguished by its distinct petal-like bracts which are reminiscent of rabbit ears. It's a compact and compact plant which typically will grow up to about 1-2 feet.

Spanish lavender is famous by its strong and sour smell and is utilized in many different products like scents and essential oils and bathing products. Also, it is a popular ingredient for cooking and utilized to enhance meals like paella as well as various other Spanish food items.

Lavandin (Lavandula x intermedia)

Lavandin is an hybrid form of lavender. It is an intermixture of English lavender as well as French lavender. It's a bigger plant that English lavender, and can grow up to about 2 feet. Lavandin is well-known for its potent aromatic scent. It is utilized in many items, such as soaps, essential oils, and candles.

Choosing the Right Variety

In selecting a particular lavender It is essential to take into consideration a variety of factors such as the type of soil, climate and the intended usage of the lavender. English lavender is the most well-known variety used in commercial production because of the high amount of oil and its sweet scent. It's ideal for colder climates, and has fertile soil.

French lavender is ideal for warm climates. It will tolerate soils that are slightly acidic. This is an excellent option for people who want to add a bit of colour to their gardens since it has huge violet blooms.

Spanish lavender is a great choice for hot and dry climates. It is tolerant of poor soil conditions. It's an excellent option for people searching for an easy-care plant that will add interest in their gardens.

Lavandin is an excellent alternative for those who want to make essential oils due to its high oil content, and it is an extremely hardy plant which can endure a wide range of soil types. It's a great choice for dry and hot environments.

Chapter 26: Growing Lavender

The lavender plant is an adaptable one which can be grown in many different soil types and climates. To ensure maximum productivity and growth it is essential to provide an appropriate conditions for the soil, climate and conditions for water. In this article we'll look at the soil, climate and essential water needs for lavender cultivation.

Soil Requirements

Lavender is best suited to fertile soil that has an acidity of between 6.5 to 7.5. It is essential to stay clear of soils that are too acidic or alkaline because this could hinder development and diminish yields. It is preferential to grow in soil high in organic matter therefore, incorporating compost and other organic material in the soil prior planting will help increase the fertility of soil.

It's also essential to make sure the soil is properly aerated since lavender plants require oxygen for their growth. In order to improve drainage and aeration, you can consider the addition of perlite or sand to soil mixes.

Climate Requirements

Lavender is a tough plant that is tolerant of a large variety of temperatures and conditions. There are different varieties that are different in their requirements for temperature which is why it's important to pick the appropriate one for your particular climate.

English lavender is suitable to colder climates. It is able to endure temperatures of as low as 20°F. French lavender is more suitable for warmer climates. It can endure temperatures up to 90degF. Spanish lavender is ideally suited to dry, hot climates. They can withstand temperatures of up to 100degF.

It is vital to know that lavender needs a time of dormancy during cold weather for it to bloom. That means it's ideal to be grown in areas with distinct seasons. This lets the plant be in a state of winter dormancy prior to spring's growth starts.

Water Requirements

Lavender is a drought-resistant plant which can withstand drought with only water. It is nevertheless crucial to make sure that your plant gets enough irrigation during the growth season for healthy growth and to maximise yields.

The first year of development, lavender needs regular irrigation to build a sturdy root system. Once the plant has been established, it will be able to live by requiring only a little water, so long as it gets enough rain. If you live in areas with dry summers, irrigation is often required for optimal development and production.

It is crucial to be cautious about the habit of overwatering your lavender because it could cause root rot, as well as other fungal infections. To avoid overwatering, make sure that your soil is adequately drained and let the soil be a little dry between each watering.

Chapter 27: Harvesting And Drying Lavender

The process of harvesting and drying lavender is a crucial stage in the process of production which determines the scent and quality of the product. In this article we'll go over how to harvest and drying lavender in order to maintain the fragrance and quality.

Harvesting Lavender

The timing for harvesting lavender is vital to assure the highest quality and quantity. It is best to harvest lavender in the early stages of buds starting to open, but before the flowers have fully blossomed. This is known by"early bloom" or "early bloom" phase and is usually observed between mid and late summer, based on nature and weather conditions.

For harvesting lavender, you will need an incredibly sharp pair of pruning shears. Cut the stems in the middle of the leaves, leaving a tiny length of the stem tied to the blooms. The stems can then be tied in bundles to allow to dry. The key is to care with the flowers in order to protect them from damage and also to avoid bruising which could affect the scent and quality of the finished item.

Drying Lavender

Following harvesting, lavender needs to be dried as quickly as possible to stop the growth of mildew and mold. The process of drying is best done in a dark, cool and ventilated space to ensure the integrity and aroma of the flower.

The most traditional method of drying lavender is to wrap the stems together into tiny bundles and then hang the bundles upside down in a dry, cool area for some months. The lavender will dry naturally in the stems, which means that the dried flowers is used to serve a range of uses such as cooking, medicinal as well as decorative.

A different method of drying lavender is using the dehydrator. Dehydrators are machines that eliminates the moisture from lavender flowers by using low temperatures as well as a fan. It is more efficient than conventional methods and is able to produce dried lavender that has a powerful aroma.

Storing Dried Lavender

Once dried, lavender must be kept in a dark, cool and dry location so that sunlight and moisture do not get from harming the flower. The lavender can be kept in many different containers like glass jars, bags of paper as well as vacuum sealed bags. It is crucial to label containers by the harvest date as well as the type of lavender in order to make sure that the flower is utilized in the earliest possible time before they begin to lose their aroma.

Chapter 28: Lavender Products

Lavender is an adaptable plant which can be utilized to make a range of items, that range from essential oils to delicious desserts. In this section we'll look at the various varieties of lavender items and ideal methods to create these.

Essential Oils

Lavender essential oil can be described as one of the most well-known essential oils widely used all over the world. It is used extensively as an aromatherapy treatment to aid relaxation and decrease anxiety. To make essential oils of lavender The flowers are then steam distillation, which removes essential oils that are fragrant from the blooms.

For the production of the highest quality lavender essential oil it is essential to select the finest lavender flowers which have been picked and dried correctly. The essential oil must be kept in a dark, cool space in airtight container for its scent and its quality.

Culinary Treats

The herb is also employed in many edible treats like syrups, teas, as well as baked goods. For culinary recipes made with lavender, you need to choose culinary-grade lavender that hasn't been treated with pesticides or any other chemical substances.

If you are cooking with lavender It is crucial to limit the amount of lavender you use because it could overwhelm other flavorings. Flowers should be taken from the stem, and then chopped into small pieces prior to adding them to dishes.

Decorative Items

It is also employed in items of decor like wreaths, sachets, or potpourri. In order to create these products dried lavender flower petals are usually employed. These flowers are mixed with dried other flowers, spices or herbs for unique and aromatic accessories.

In the process of creating ornaments using lavender, it's important to take care when handling the flower in order to prevent damage, and also to be sure that they maintain their scent. It is recommended to store the items in a dark, cool and dry location for them to keep their beauty and scent.

Other Lavender Products

The scent of lavender is used in various different products like soaps, lotions as well as candles. They typically make use of lavender essential oils or dried lavender flowers for their scent.

In the production of lavender-based products, it's important to make use of high-quality ingredients as well as adhere to the best practices when making the item. They should be kept in a dry, cool and dark area for them to keep their freshness as well as their scent.

Chapter 29: Marketing Your Lavender Business

After you've grown lavender and developed a selection of items and products, it's time to think about marketing your company. In this section we'll look at how to build a powerful image for your brand, develop effective sales strategies and market your lavender-related products in the most effective way to reach out to the appropriate target market.

Branding Your Lavender Business

An effective brand image is essential to the growth for your lavender company. Your brand must reflect your mission, values, and also the distinct characteristics of your products made from lavender. Here are a few suggestions for creating an identity for your brand:

Select a name that is memorable for your business, one that is a reflection of the essentials of lavender.

Create a logo that is memorable, simple and visually attractive. Think about using colors such as lavender, violet, green and white to create your logo.

Create a brand voice which reflects your personality as a business. Does your voice reflect you? fun, serious or something in-between?

Create a story for your brand which connects your target market and is a story about your business in the field of lavender. The story of your brand should emphasize the mission, values as well as what makes your products in the field of lavender above the rest.

Sales Strategies for Lavender Products

After you've created a an identity for your brand It's time to concentrate on selling strategies to help market your products. Below are some strategies for selling you should consider:

Sell directly to customers Think about selling your lavender-related products at market stalls, crafts fairs, or other events in which your ideal customer will likely be.

Partner with retailers: Contact merchants who may be looking to carry the lavender items in their shops. These could be specialty stores as well as health food stores as well as gift stores.

Consider selling your lavender items on your site or via e-commerce platforms such as Etsy and Amazon.

Promotion Strategies for Lavender Products

The promotion of lavender products is crucial in establishing awareness and boosting sales. These are some marketing strategies to think about:

Social media: Make use of social media websites including Facebook, Instagram, and Twitter to advertise the products you sell in your lavender garden. Upload photos of your lavender products and behind-the scenes photos of your lavender farm and news about events coming up.

Chapter 30: The Wellness Benefits Of Lavender

It is a well-known plant as a relaxing and aromatic properties. This makes it a favorite option for relaxation and aromatherapy. However, did you realize that lavender has numerous health advantages? In this article we'll look at the numerous health benefits associated with lavender, and the best ways to include it in your daily routine to enhance your general health.

Stress Relief

The most widely known positive effects associated with lavender is the capability to ease anxiety and stress. Research has shown that inhaling the aroma of lavender may aid in lowering levels of cortisol hormone that causes stress, leading to a sense of peace and peace. It is possible to incorporate lavender into your stress relieving routine with lavender essential oils in diffusers, taking an energizing bath in lavender, and using products that have lavender fragrances, such as candles and lotions.

Improved Sleep

Along with reducing anxiety, lavender is also known to aid in improving sleep quality. Research has shown that inhaling the lavender scent prior to going you go to bed could improve your sleep quality and length, providing a natural sleep helper. It is possible to incorporate lavender in your sleep routine with lavender essential oils in diffusers, including lavender in your bath and using products that smell of lavender such as pillows or sleep masks.

Headache Relief

It has also been proven to have a positive effect on reducing the severity and frequency of headaches. The study in European Journal of Neurology found that inhaling the aroma of lavender for fifteen minutes could help decrease the intensity of migraine headaches. It is possible for headache relief through inhaling essential oils of lavender or applying a balm infused with lavender on your temples or by using a lavender-scented eye pillow.

Skin Care

The lavender plant also has a variety of skin benefits. The antibacterial and anti-inflammatory characteristics make it an ideal natural treatment to treat eczema and acne and various other skin problems. It can also calm and soothe damaged or dry skin. It is possible to incorporate lavender into your daily skin care routine through lavender-infused cleansing products or toners. You can also use them as moisturizers as well as applying the lavender essential oils directly onto the skin.

Pain Relief

Lavender is used throughout history as a cure for pain. Its anti-inflammatory properties help in reducing inflammation and swelling and is a fantastic remedy for ailments such as muscular pain and arthritis. It is possible to use lavender oil to reduce discomfort using lavender essential oil directly on the area affected by using lavender-scented heat cold packs. Or, you can use the massage oil infused with lavender.

Chapter 31: Lavender As A Home Decorator To Create Soothing Spaces The Scent Of This Flower

It's not just many positive health effects, but can also provide a relaxing and soothing ambience in your house. In this section we'll look at the numerous ways to include lavender in your decor to make a relaxing and tranquil area.

Lavender Color Palette

One of the easiest methods to include lavender in your decor for home is employing a palette of lavender colors. It is a soothing and relaxing color that is able to be used to create an accent color, or the dominant color for the room. It is possible to paint with lavender on the walls of your home, or even lavender accents, such as curtains, pillows, or blankets for an accent of color to the decor.

Lavender Plants

The lavender plants are not just gorgeous, they provide a soothing and soothing scent. It is possible to incorporate lavender plants in your decor for the home by putting the plants in a sun-lit area inside your home for example, a window or the balcony. Also, you can use the lavender plants for the outdoor décor, for example, on your deck or even in your garden. Lavender's scent plant may help reduce anxiety and stress, as well as create the perfect atmosphere for relaxation and peace at home.

Lavender Candles

Lavender candles have a soft and relaxing scent that could assist in creating a peaceful environment in your home. Candles made of lavender can be used in your living space, or any room you'd like to create a serene place. The gentle glow of candles, paired with the relaxing aroma of lavender will aid in creating a relaxing and relaxing environment.

Lavender Pillow Mist

Lavender mist on your pillows is an excellent method to bring the relaxing scent of lavender in the decor of your bedroom. The lavender mist for your pillow to brighten the look of your linens and pillows and spray it into the air to create a calm and peaceful atmosphere. The lavender mist in your pillow can assist in reducing stress and anxiety and enhance sleep quality. This makes it the perfect alternative for the bedtime routine.

Lavender Linens

Lavender-scented linens add an element of luxury to the decor of your home while giving you the soothing and relaxing advantages of lavender. Lavender-scented pillows, sheets, and bath towels in your bedrooms as well as in your bathroom to create an atmosphere of relaxation. The delicate scent of lavender will help reduce stress and anxiety as well as create the perfect atmosphere of calm and peace within your house.

Chapter 32: Advanced Lavender Techniques

In this article we'll look at some advanced methods for cultivating lavender. These techniques can help you enhance your yield and the overall health of the lavender plants. The techniques covered include pruning, propagation, as well as maximizing yield.

Propagation

Propagation refers to the method to create new plants of lavender by combining existing plants. There are two primary ways of propagation: either by cuttings or seeds. Growing lavender from seeds could be a lengthy time because the seeds require about a week to sprout and can require several years of maturation. Growing by cuttings is more efficient, since the plants will be harvested within just a few months.

For the propagation of lavender through cuttings, you must follow these instructions:

Choose a healthful and mature lavender plant, which has robust stems and healthy foliage.

Cut off a stem from the plant, which measures approximately 4-6 inches in length and contains multiple leaf sets.

Take the leaves off the base of the stem. This will leave only a few leaves at the upper.

Dip the cut side of the stem with roots-forming hormone.

Put the stem into the pot in a container that is well-drained.

Keep the soil damp and put it in a sun-lit spot.

Within a couple of months, the plant is expected to have rooted and should be in good shape to be transplanted into a larger pot, or directly into the soil.

Pruning

Pruning is a crucial aspect of cultivating lavender since it assists in keeping the plants productive and healthy. Pruning is also a way to stop the plant from becoming heavy and sluggish. The ideal time for pruning lavender is early in spring when spring is bringing new growth.

To trim lavender, do these steps:

Use a pair cutting shears that are sharp and clean.

Cut the stems back in a third-to-one-half increment of the length. Leave fresh growth in the lower part and the top.

Take out branches that are damaged or dead.

The plant should be pruned into an oval or round shape with a few inches space between the plants, allowing air circulation.

Maximizing Yield

In order to maximize the production of the lavender plants you have There are many ways to go about it:

Pick the appropriate type: Certain lavender cultivars are more productive than others. It's essential to pick the appropriate one for your particular growing environment.

It is important to plant in the correct place: The lavender needs sunshine and a well-drained soil in order for it to flourish, so make certain to plant it in an area that fulfills the requirements.

Fertilize frequently the lavender needs a well-balanced fertilizer in order to thrive so make sure you fertilize it regularly throughout the growing time.

Water correctly: It is essential to water your lavender regularly however, excessive watering can cause root decay. Make sure to water your lavender thoroughly each week, and let that the dirt to dry between each watering.

Harvest at the appropriate time you should harvest your lavender once the flowers are in bloom, but before they start to turn brown. It is typically at the beginning of the morning when the sun has not been able to dry the blooms.

www.ingramcontent.com/pod-product-compliance
Lightning Source LLC
Chambersburg PA
CBHW071443080526
44587CB00014B/1974